Diagnostic Manual for the Identification of Insect Pathogens

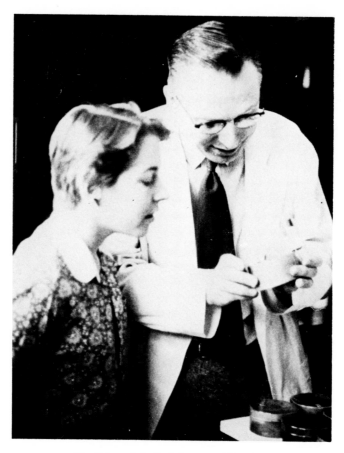

Dr. Edward A. Steinhaus and Student

Diagnostic Manual for the Identification of Insect Pathogens

George O. Poinar, Jr.
and Gerard M. Thomas

University of California at Berkeley

PLENUM PRESS · NEW YORK AND LONDON

Library of Congress Cataloging in Publication Data

Poinar, George O
   Diagnostic manual for the identification of insect pathogens.

   Bibliography: p.
   Includex index.
   1. Insects – Diseases – Diagnosis. 2. Parasites – Insects – Identification. 3.
Micro-organisms Pathogenic – Identification. I. Thomas, Gerard M., joint author.
II. Title.
SB942.P64   595.7'02   77-15977
ISBN 0-306-31097-X

First Printing – February 1978
Second Printing – April 1982

© 1978 Plenum Press, New York
A Division of Plenum Publishing Corporation
233 Spring Street, New York, N.Y. 10013

Printed in the United States of America

To the memory of
EDWARD A. STEINHAUS
who established the first Insect Disease Diagnosis Laboratory
in 1944 at the University of California at Berkeley

PREFACE

This manual was prepared for the diagnosis of insect diseases caused by infectious agents. The agents (or pathogens) included here are fungi, protozoans, bacteria, viruses, and rickettsias.

The present work was prepared after much deliberation and discussion with students and teachers who felt a guide of this type would be valuable for diagnosing the microbial diseases of insects. It was modeled after a seminar given on the same subject at Berkeley, which had as its major goal the recognition and identification of insect pathogens for practical purposes. The present work includes numerous timesaving "short cuts" which were developed after years of experience of diagnosing insect diseases.

Although emphasis is placed on *identification,* general background information on the various pathogens is also included. Thus, under each of the five groups of pathogens, the following topics are discussed: (1) various types of associations with insects, (2) definition and classification, (3) general life cycle, (4) characteristics of diseased insects, (5) factors affecting natural infections, (6) methods of examination, (7) isolation and cultivation, (8) important taxonomic characters, (9) tests for infectivity, (10) storage, (11) an illustrated key to the genera (or group in the case of viruses), and (12) literature, especially that pertaining to identification.

Although often included with insect pathogens, entomogenous nematodes are not covered here since illustrated keys to those genera that infect insects are already available (Poinar, 1975, 1977).

Techniques for diagnosing insect diseases already have been

vii

presented by one of us (Thomas, 1974), and a glossary containing terms used in invertebrate pathology has been prepared by Steinhaus and Martignoni (1970). Other terms used in this manual will be found in most textbooks covering the various pathogen groups.

Many of the photographs presented in this manual were made from material submitted to the Insect Pathology Diagnostic Laboratory at Berkeley, California. This diagnostic laboratory was established in 1944 by the late Professor Edward A. Steinhaus and has been continued up to the present. The results of this service have been published at regular intervals (Steinhaus, 1951) (Steinhaus and Marsh, 1962) (Thomas and Poinar, 1973). Other photographs were supplied by specialists, who are cited in another section of this work. We are deeply indebted to the above for helping us to make the manual complete.

We hope this work will serve its intended purpose and welcome any comments which would make it more useful.

Berkeley

George O. Poinar, Jr.
Gerard M. Thomas

ACKNOWLEDGMENTS

We are very grateful to the following persons who supplied us with pathogens or photographs of pathogens unavailable to us: W. Balamuth, L. P. Brinton, W. Burgdorfer, H. Chapman, J. Couch, B. F. Eldridge, B. Federici, D. Forgach, T. Fukuda, R. Goodwin, E. Greiner, E. Hazard, Roberta Hess, Darlene F. Hoffmann, J. Hurlimann, A. M. Huger, M. Laird, A. Kaplan, W. R. Kellen, M. Martignoni, E. M. McCray, Jr., B. Nelson, D. Sanders, Ruth Sluss, Clara M. Splittstoesser, and J. Weiser.

We would also like to thank the following persons for critically reviewing various sections of the manual: J. V. Bell, I. M. Hall, W. R. Kellen, D. E. Pinnock and Y. Tanada. We are also grateful to Mrs. E. A. Steinhaus for supplying us with photographs of Dr. E. A. Steinhaus.

Finally, we would not have attempted this project without the enthusiastic support of many students who attended our seminars on the diagnosis of insect diseases. They, indirectly, had a role in the preparation of this manual.

CONTENTS

IDENTIFICATION OF THE GROUPS OF INSECT PATHOGENS

Disease may be considered to be a state in which the body, or a tissue or organ of the body, is disturbed either functionally or structurally or both. Insect diseases may be infectious, that is, caused by pathogens, or noninfectious—caused by abiotic factors (temperature, starvation, chemicals, injury). As only the former condition is dealt with here, one should keep in mind that a seemingly diseased insect without any visible foreign particles may have a noninfectious disease, and, aside from the possibility of a noninclusion virus, factors other than those pathogens listed here should be considered as causative.

The following key serves as a general guide for separating the pathogen groups. It should not be regarded as infallible since it is sometimes difficult to distinguish between certain protozoan and fungal "spores" or the cells of a rickettsia and the capsules of a granulosis virus. Fortunately, in many cases, the pathogen group will be obvious. If not, the following key should be helpful.

1

KEY TO THE GROUPS OF INSECT PATHOGENS*

1. Insects covered with mycelium (usually white, yellow, or green) or elongate fruiting structures bearing spores; cadaver often mummified, hard or cheeselike in consistency; tissues containing hyphae—Fungi
1. Insects not covered with mycelium or elongate fruiting structures; tissues not containing hyphae; particles present in hemocoel—2
2. Particles spherical, stain reddish with Sudan III (Figure 1) (see section on Techniques)—Fat globules
2. Particles variable in shape, do not stain reddish with Sudan III—3
3. Particles showing birefringence under polarized light (Figure 2)—Urate or other types of crystals
3. Particles not showing birefringence under polarized light—4
4. Particles generally motile and rod-shaped (occasionally spherical and nonmotile); develop in the hemocoel (rarely only in the intestine); spores may be present—Bacteria
4. Particles generally nonmotile and rarely rod-shaped (except Rickettsia); develop in host cells or tissues (some ciliate protozoans are motile and multiply in the hemocoel of their hosts); infectious particles (spores, sporozoites, etc.) or structures containing infectious particles (oocysts, etc.) variable in size and shape—5
5. Infectious particles rod-shaped; just visible under the light microscope—Rickettsia
5. Infectious particles (spores, sporozoites, etc.) or structures containing infectious particles (cysts, etc.) rod-shaped or spherical; generally easily visible under the light microscope—6
6. Small to minute particles (greatest diameter generally less than 10.0 μm) consisting of polyhedra and capsules which usually dissolve in a weak solution of NaOH—Inclusion viruses
6. Particles of various shapes (diameter generally ranges from 2.0

---

*With stages visible under the light microscope. The absence of infectious particles in an apparently sick insect may indicate that the disease is noninfectious or caused by a noninclusion virus.

to 20.0 μm) which do not dissolve in a weak solution of NaOH—7

7. Infectious particles (spores, sporozoites, etc.) or structures containing infectious particles (cysts, etc.) generally formed within host tissues; some ciliated forms multiply in the hosts' hemolymph—Protozoa

7. Infectious particles (spores) or structures containing infectious particles (cysts) generally formed in the hosts' hemolymph; cilia absent—Fungi

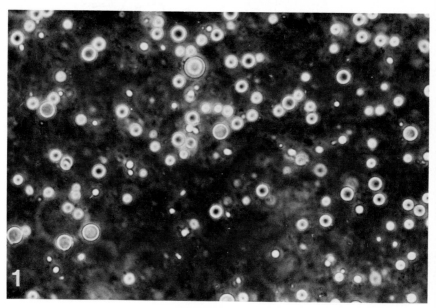

FIGURE 1

Fat globules from the hemocoel of a healthy insect. ×640

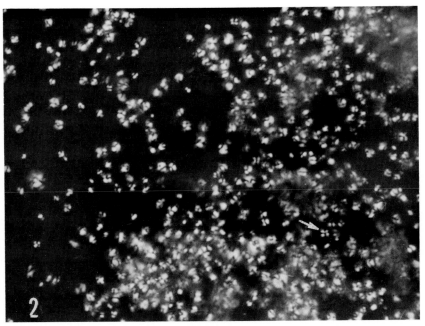

FIGURE 2

Urate crystals under polarized light. Note cross (arrow). ×640

IDENTIFICATION

One of the most difficult aspects of fungal identification is the preparation of slides clearly showing the diagnostic characters of the fungus. Characteristics to be noted are the size and shape of the spores, their attachment to the hyphae, and the presence or absence of hyphal septa and clamp connections. Is the mycelium modified into a stroma or synnemata? Are the spores motile, septate, catenulate, or borne in slime drops?

Lastly, many fungi are specific to certain hosts or host groups and every attempt should be made to identify the diseased insect.

TESTING FOR PATHOGENICITY

In testing the pathogenicity of an unknown fungus, it is best to use insects of the same species originally attacked. If not available, then laboratory test insects such as larvae of *Galleria mellonella* can be used (see chapter on Techniques for methods of rearing). Since most fungal pathogens enter through the host's cuticle, it is usually sufficient to place spores directly on the insects' integuments with a sterile instrument; or the insects can be made to walk over a sporulating fungus culture. The treated insects should then be kept in a moist, warm environment, which induces spore germination.

Spores of fungi which invade through the intestine can be collected after washing the culture surface with sterile water and then mixed with the host's food. They can also be introduced into the insect's buccal cavity with a glass-tipped hypodermic syringe.

STORAGE

Many fungi can be stored for years under lyophilization, liquid nitrogen, or on silica gel [see Bell and Hamalle (1974) for a discussion of the latter process]. More sensitive forms can be maintained on agar slants capped with wax under refrigeration or immersed in mineral oil at room temperature. Of course, periodic transfers must be made.

LITERATURE

For general taxonomic studies on fungi, the treatises of Ainsworth *et al.* (1973) and Barnett and Hunter (1972) are useful. Two recent books on entomogenous fungi have been written by Evlakhova (1974) and Koval (1974). For host–pathogen relationships, Steinhaus's *Advanced Treatise of Insect Pathology* (1963) and Madelin's review article (1966) can be consulted. Roberts and Yendol (1971) discuss the use of fungi for insect control, and Bell (1974) presents a general review of insect mycoses. Ainsworth and Bisby's *Dictionary of the Fungi* (Ainsworth, 1961) gives an excellent account of mycological terms and definitions of genera and higher fungal taxons.

KEY TO COMMON GENERA

1. Mycelium nonseptate (coenocytic), variable in development; motile cells may be present (Mastigomycotina)—**2**
1. Mycelium septate, usually well developed; motile cells absent—**7**
2. Mycelium usually well developed; sexual reproduction results in the formation of nonmotile rounded zygospores; asexual reproduction results in the formation of sporangia or conidia which are produced inside or on the exterior surface of the host (Zygomycetes)—**3**
2. Mycelium usually sparse; reproduction results in the formation of motile zoospores; thick- or thin-walled resting spores or sporangia often present inside the host—**4**
3. Conidia forcibly discharged, produced outside the host— *Entomophthora* Fres. (Figures 3, 4, 5). For discussions of this genus, see MacLeod (1963) and Waterhouse (1973). *Strongwellsea* Batko and Weiser and *Tarichium* Cohn are doubtful genera and may represent aberrant species of *Entomophthora* [see Waterhouse (1973)].
3. Conidia not forcibly discharged, produced inside the host (in abdominal cavities)—*Massospora* Peck. (Figure 6). For a discussion of this genus, see MacLeod (1963) and Speare (1921).

4. Hyphae unwalled, become converted into resting spores or sporangia (usually sculptured and pigmented in *Coelomomyces*); kidney-shaped, or round zoospores posteriorly uniflagellate (Chytridiomycetes)—**5**

4. Hyphae walled; round zoospores biflagellate, formed in a vesicle at the tip of a discharge tube (Oomycetes)—**6**

5. Attack aquatic insects, especially mosquito larvae; host turns yellow, orange, or brown due to the color of the mature resting spores—*Coelomomyces* Keilin (Figure 7). For an account of this genus, see Couch and Umphlett (1963); Whistler *et al.* (1974) has shown copepods to be alternate hosts of *Coelomomyces* pathogenic to mosquitoes. The related genus *Coelomycidium* Debaiseux (Figure 8) will also key out here. This pathogen has been recovered from black flies (Simuliidae) and often turns the host a pinkish hue.

5. Attack terrestrial insects; transforms host tissues into an orange-colored mass of globular resting spores or sporangia which become powdery when dry—*Myiophagus* Thaxt. (= *Myrophagus*). For an account of this genus, see Karling (1948).

6. Hyphae become broken up into segments, each of which gives rise to distinct sporangia or gametangia—*Lagenidium* Schenk (Figure 9). [For an account of this genus, see Umphlett and Huang (1972).]

6. Hyphae mostly vegetative, only certain portions giving rise to sporangia or gametangia—*Pythium* Pringsh. (Figure 10). Clark *et al.* (1966) have shown that members of this genus can be potential pathogens, similar to *Saprolegnia,* another member of the Oomycetes that has been studied by Rioux and Achard (1956).

7. Sexual reproduction results in the formation of transversely septate basidia which bear 4 basidiospores. Parasites of scale insects; stroma flat, appressed to bark of tree (Septobasidiales) —**8**

7. Sexual reproduction absent or resulting in the formation of an ascus containing 8 ascospores. Asexual reproduction with conidia borne on conidiophores—**9**

8. Epibasidium arises from a hypobasidium (probasidium);

parasitizes entire colonies of scale insects—*Septobasidium* Pat. (Figure 11). For an account of this genus, see Couch (1938).

8. Epibasidium arises from an elongated uredospore; parasitizes single scale insects—*Uredinella* Couch. See Couch (1937) for an account of this genus.

9. Sexual stage resulting in an ascus containing 8 ascospores (Ascomycotina)—**10**

9. Sexual stage absent, hyphae and conidia present; mycelium generally found inside and on the surface of the host (Deuteromycotina)—**14**

10. Mycelium scanty or lacking; plants unicellular, reproducing asexually by budding, fission, or both. When produced, ascospores are borne in a naked ascus—yeasts of the class Hemiascomycetes. Some genera in this class are insect pathogens and many others are associated with healthy insects. *Candida* Berkout (Figure 12) can be pathogenic to insects (see Martignoni *et al.,* 1969). Other yeasts in the genera, *Mycoderma, Saccharomyces,* and *Blastodendrion* have been isolated from insects. However, their pathogenicity has not been elucidated; see Steinhaus (1949) and Miller and van Uden (1970) for coverage of these forms.

10. Mycelium generally present; ascospores borne in ascocarps (fruiting bodies); asexual reproduction, when present, by conidia; budding rare or absent—**11**

11. Mycelium sparse, small bush- or hairlike growths on the surface of the insect's cuticle; thallus usually consisting of a few cells; usually only 4 ascospores produced (rarely 8); ascospores once septate—Laboulbeniomycetes (Figure 13). These fungi are generally not pathogenic and are more curiosities than anything else. However, *Hesperomyces virescens* was reported to be pathogenic to coccinellid beetles (Kamburov *et al.,* 1967).

11. Mycelium extensive; found inside as well as on the external surface of insects; asci with 8 (rarely 2–4) ascospores (or many ascospores occur in a spore ball); ascospores nonseptate or with 2 or more septa—**12**

12. Ascospores borne in spore balls in dark-colored cysts which appear as tiny black specks on mummified bee larvae; asco-

spores nonseptate—*Ascosphaera* Olive and Spiltoir (Figures 14, 15). Species in this genus cause chalk-brood disease in honey and leafcutter bees. See Skou (1972) for a review of this genus and related species.

12. Ascospores not borne in spore balls in cysts; ascospores with 2 or more septa—**13**

13. Parasites of scale insects; mycelium forms a black cushion-shaped mat covering one or more scales; asci with thick dark septate ascospores embedded in the fungus stroma—*Myriangium* Mont. and Berk. (Figure 16). See Miller (1940) for a discussion of this genus.

13. Parasites of various terrestrial insects (rarely scales); the host is filled with septate mycelium which forms an aerial stroma extending out of the insect's body for some distance; perithecia containing cylindrical asci occur on the fertile portion of the stroma; ascospores filiform and multiseptate—*Cordyceps* (Fr.) (Figures 17, 18). See McEwen (1963) for a discussion of this genus.

14. Hyaline, fusoid conidia born in pycnidia formed in a cavity of the fungal stroma; pycnidia usually brightly colored; parasites of white flies and scales—*Aschersonia* Mont. (Figure 19). A monograph of this genus was prepared by Petch (1921), and Mains (1959) discussed the North American species parasitic on white flies.

14. Conidia variable, not produced within a pycnidium embedded in a fungal stroma—**15**

15. Conidiophores united into synnemata or elongated hornlike structures arising from the dead insect, bearing a superficial resemblance to *Cordyceps*—**16**

15. Conidiophores variable, but not united into synnemata—**21**

16. Phialides (bottle-shaped structures which give rise to spores) borne singly on the conidiophores (Figure 20)—**17**

16. Phialides borne in groups or clusters on the conidiophores (Figure 21)—**19**

17. Phialides usually short and thick (sometimes nearly spherical); conidia dry; synnemata thick, whitish, usually not branched—*Isaria* Pers. (Figures 22, 23). May be an imperfect stage of

Cordyceps species (McEwen, 1963). See DeHoog (1972) for a recent revision of this genus.

17. Phialides elongated, slender, conidia covered with mucus; synnemata slender, buff colored, usually branched—**18**

18. Phialides enlarged at base; conidia not in heads—*Hirsutella* Pat. (Figures 24, 25, 26). May be an imperfect stage of *Cordyceps* species (McEwen, 1963). See Mains (1951) for a review of the genus.

18. Phialides not enlarged at base; conidia in heads—*Synnematium* Speare (Figures 27, 28). See Mains (1951) for a review of this genus.

19. Phialides obtuse at apex (Figure 20); conidia borne singly (Figure 20), not catenulate; synnemata cylindrical—*Hymenostilbe* Petch (Figure 29). May be an imperfect stage of *Cordyceps* species (McEwen, 1963).

19. Phialides pointed at apex; conidia borne in chains (Figure 21)—**20**

20. Tips of synnemata swollen—*Insecticola* Mains. See Mains (1950) for a discussion of this genus.

20. Tips of synnemata not swollen, but cylindrical or pointed—*Akanthomyces* Leb. May be an imperfect stage of *Cordyceps* species (McEwen, 1963). See Mains (1950) for a discussion of this genus.

21. Conidiophores covering a cushion-shaped stroma; conidia dark; parasitic on scale insects—*Aegerita* Pers. For the biology of this genus, see Morrill and Black (1912).

21. Conidiophores borne over the surface of the host, not arising from a stroma covering the host; conidia usually hyaline; may or may not parasitize scale insects—**22**

22. Two types of conidia present; slender, usually septate, canoe-shaped macroconidia and smaller microconidia (may not occur together); conidia not borne in chains—*Fusarium* Link (Figures 30, 31). For notes on the pathogenesis of species, see Madelin (1963) and Hassan and Vago (1972).

22. Typically only one type of conidia present—**23**

23. Conidia borne in slime or mucus balls at the tips of phialides (especially in older, mature mycelium)—*Cephalosporium*

Corda. May be an imperfect stage of *Cordyceps* species (McEwen, 1963). For relationships with insects, see Ganhão (1956).

23. Condida not collecting in slime drops at the apex of the phialides—**24**

24. Conidia borne singly, not catenulate; fertile portion of conidiophore zig-zag in shape and drawn out at tip—*Beauveria* Vuill. (Figures 32, 33). See DeHoog (1972) for a revision of this genus.

24. Conidia borne in chains (catenulate)—**25**

25. Phialides borne in clusters or groups on the often enlarged apex of single conidiophores—*Aspergillus* Mich. (Figures 34, 35). For notes on the pathogenicity of this genus, see Madelin (1963) and Sussman (1951).

25. Not as above—**26**

26. Conidiophores in compact columns or groups; conidia elongate or rod-shaped—*Metarrhizium* Sorok. (Figures 36, 37). The common species *M. anisopliae* is a characteristic green color. For pathogenicity studies, see Zacharuk and Tinline (1968).

26. Conidiophores not in compact groups, conidia variable in shape, usually globose—**27**

27. Phialides in nondiverging clusters, usually green—*Penicillium* Link (Figure 38). For pathogenicity studies see Sen *et al.* (1970).

27. Phialides divergent in loose groupings—**28**

28. Usually white, yellow, rose or red colonies—*Paecilomyces* Bain (Figure 39). See Brown and Smith (1957) for a taxonomic discussion of this genus.

28. Green colonies—*Nomuraea* Maublanc (= *Spicaria* Harting) (Figures 40, 41). May be an imperfect stage of *Cordyceps* species (McEwen, 1963).

FIGURE 3

The trypetid fly, *Anastrepha suspensa*, infected with *Entomophthora echinospor*
×5

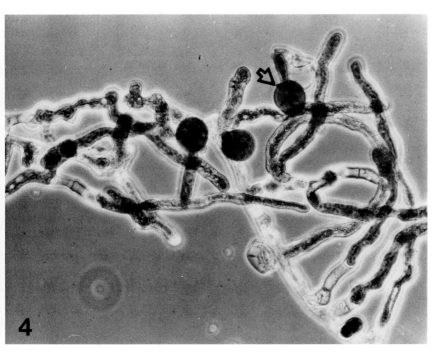

FIGURE 4

Mycelium and spores (arrow) of *Entomophthora virulenta.*   ×213

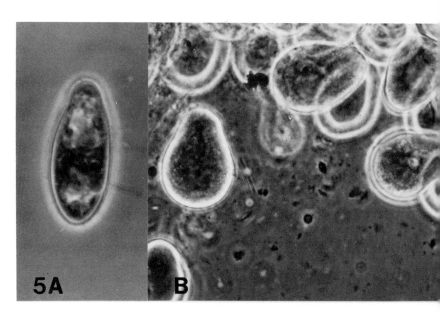

FIGURE 5

Spore(s) of *Entomophthora* sp. (near *aphidis*) (A) and *E. grylli* (B). ×720

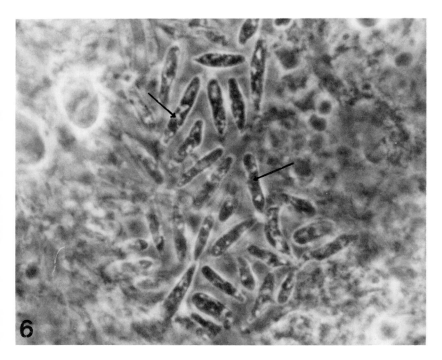

FIGURE 6

Conidia (arrows) of *Massospora* sp. in the abdominal cavity of a cicada. ×2000

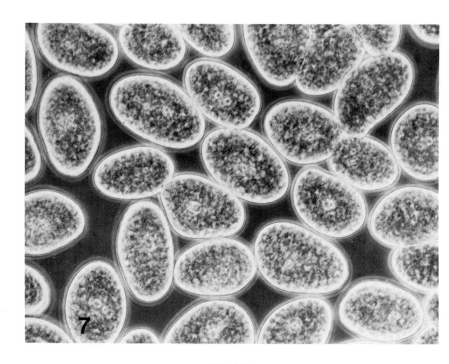

FIGURE 7

Resting spores of *Coelomomyces* sp. from a chironomid larva. ×600

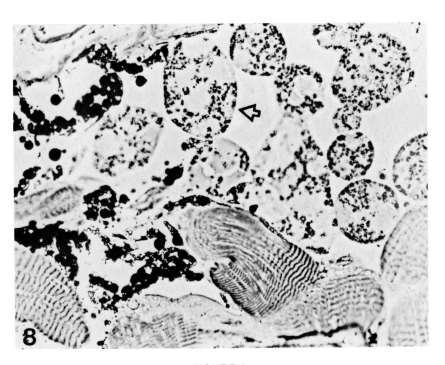

FIGURE 8

Developing sporangia of *Coelomycidium* sp. (arrow) from *Simulium vittatum* (courtesy of Brian Federici).   ×700

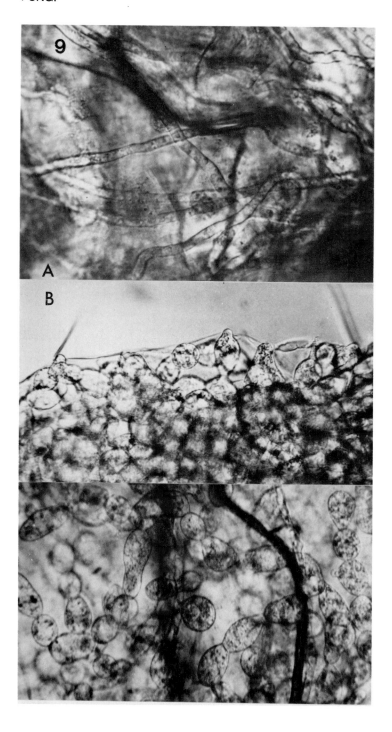

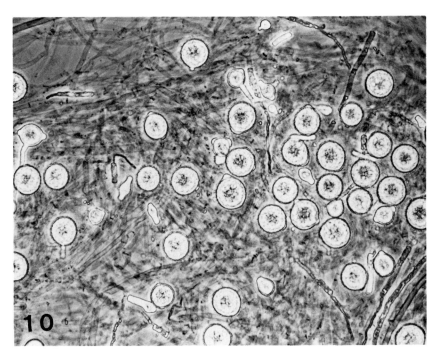

FIGURE 10

Sporangia of *Pythium* sp. (courtesy of J. Hurlimann).   ×225

FIGURE 9

Lagenidium giganteum in *Culex pipiens quinquefasciatus;* (A) hyphae,
(B) developing sporangia (courtesy of Elmo M. McCray, Jr.).   ×500,
reproduced @ 85%.

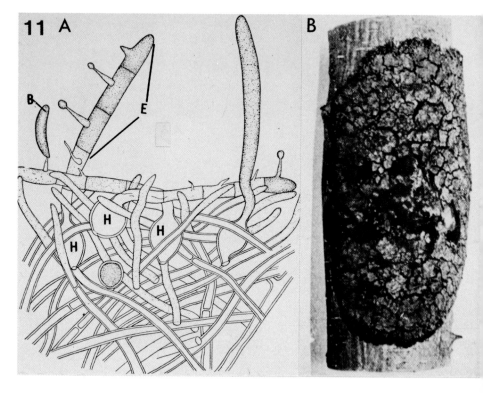

FIGURE 11

Septobasidium sp. (A) Mycelium and basidium (E, epibasidium; H, hypobasidium; B, basidic
spore) (×940). (B) Fungal stroma appressed to tree (after Couch, 1938) (×1.35). A and
reproduced @ 90%.

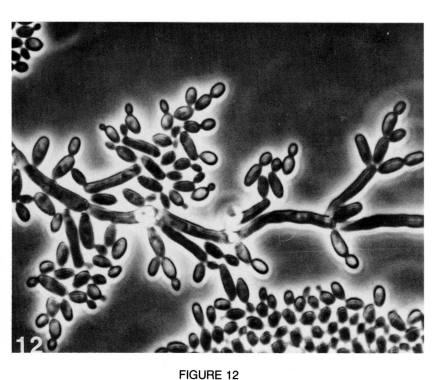

FIGURE 12

Candida sp. from a larva of the Douglas fir tussock moth
(courtesy of M. Martignoni). ×1000

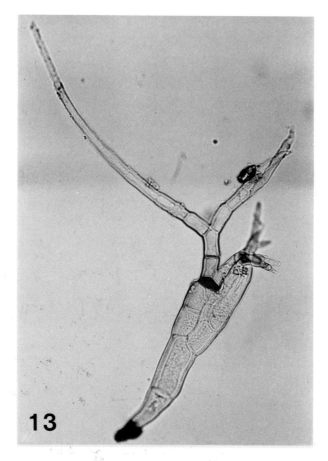

FIGURE 13

A member of the Laboulbeniales (courtesy of A. Kaplan).
×250

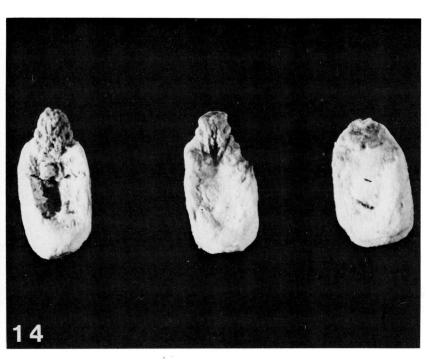

FIGURE 14

Mummified larvae of *Apis mellifera* covered with mycelium of *Ascosphaera apis.*
×6

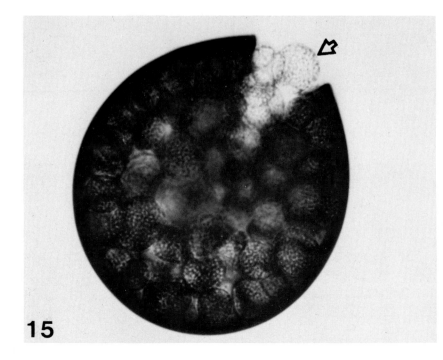

FIGURE 15

Cyst of *Ascosphaera apis* containing spore balls (arrow) filled with ascospores ×800

FIGURE 16

(A) Mycelial-mat of *Myriangium* sp. ×8. (B) Ascospores of *Myriangium* sp. ×700, reproduced @ 95%.

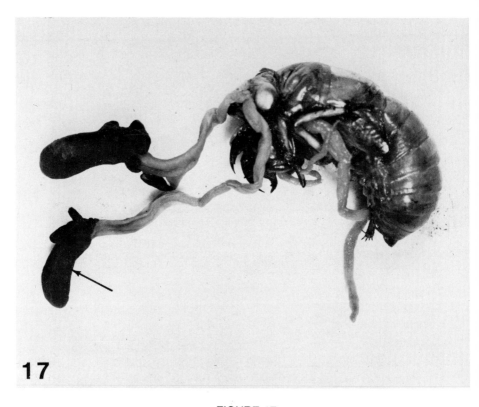

FIGURE 17

A nymph of the cicada *Diceroprocta apache* bearing aerial stroma terminated with perithecia (arrow) of *Cordyceps* sp. (Probably *sobolifera*). ×3.5

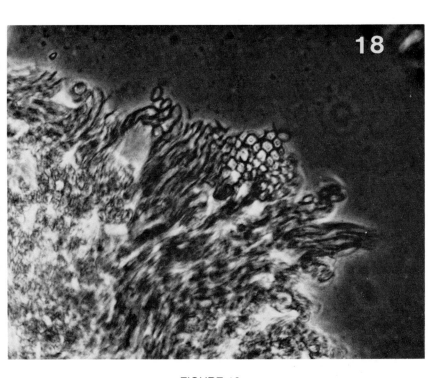

FIGURE 18

Asci of *Cordyceps* sp. containing ascospores. ×980

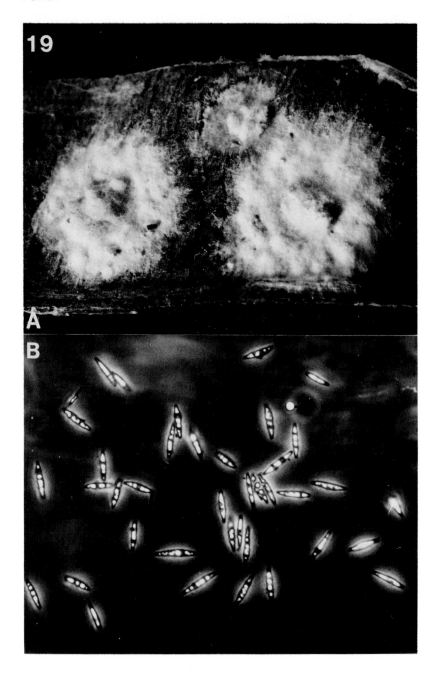

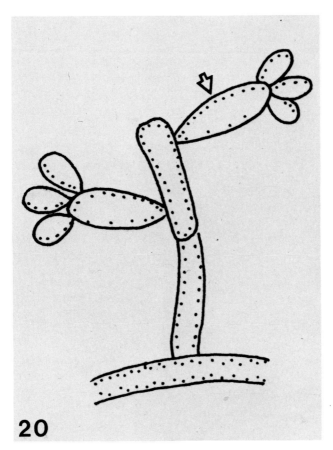

20

FIGURE 20

Phialides (arrow) born singly on the conidiophores.

FIGURE 19

Fungal stroma of *Aschersonia* sp. growing on a diaspid scale. (A) ×8
(B) spores ×770, reproduced @ 95%.

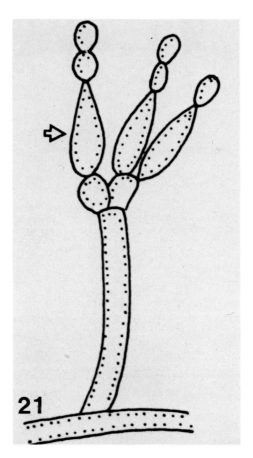

FIGURE 21

Phialides (arrow) born in groups on the conidiophores.

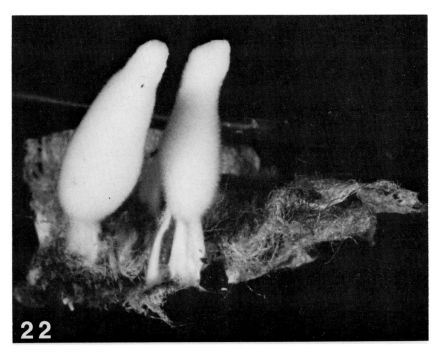

FIGURE 22

Synnemata of *Isaria* sp. emerging from a pupa of the codling moth. ×20

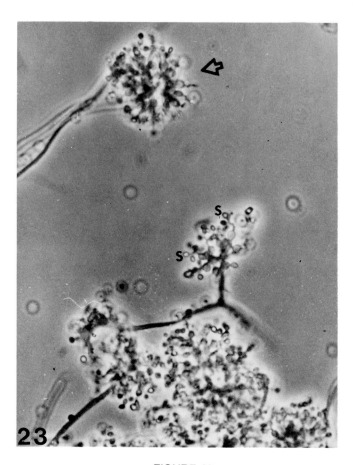

FIGURE 23

Clusters of phialides (arrow) of *Isaria* sp. grown in agar. Note spores (S). ×850

FIGURE 24

Hirsutella saussurei infecting the wasp, *Polistes olivaceous.* ×4

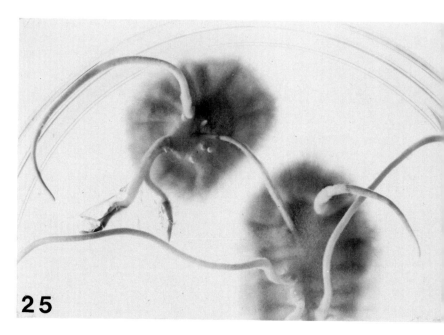

25

FIGURE 25

An agar culture of *Hirsutella* sp. showing synnemata. ×20

FIGURE 26

Phialides (arrows) of *Hirsutella* sp. ×700

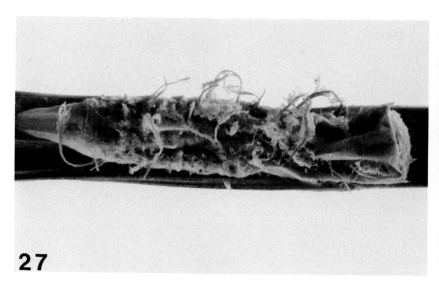

FIGURE 27

Synnemata of *Synnematium* sp. on *Panoquina* sp. ×4

FIGURE 28

Phialides (arrows) of *Synnematium* sp. on *Antiteuchus tripterus*. ×720

FIGURE 29

Hymenostilbe sp. on a dragonfly, showing the cylindrical synnemata, reproduced @ 90%.

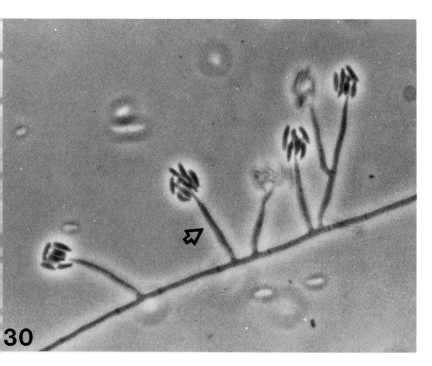

FIGURE 30

Microconidia of *Fusarium* sp. borne on phialides (arrow). ×800

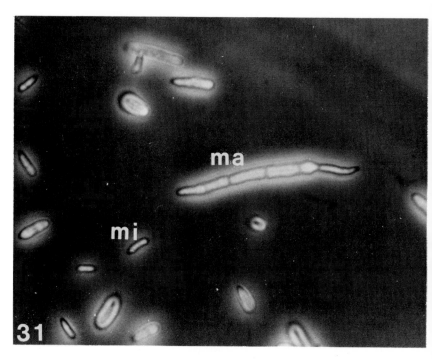

FIGURE 31

Microconidia (mi) and macroconidia (ma) of *Fusarium* sp. ×1024

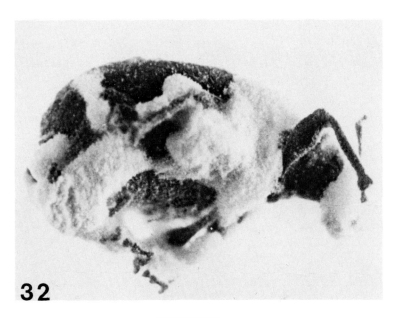

FIGURE 32

Beauveria bassiana on the weevil, *Nemocestus incomptus.* ×12

FIGURE 33

Conidia and phialides (arrows) of *Beauveria bassiana.* ×800

FIGURE 34

Aspergillus flavus (arrow) on the ant, *Atta texana.* ×12

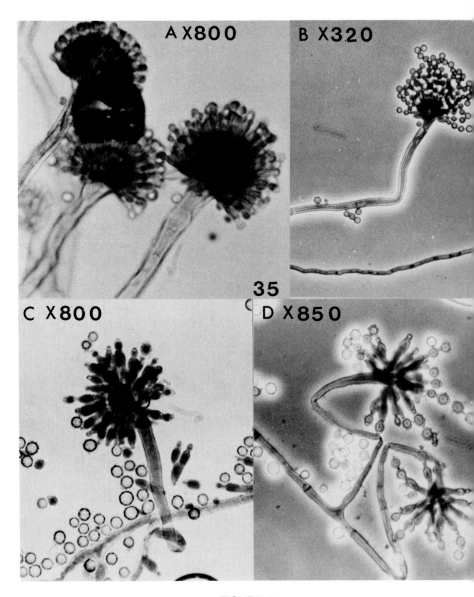

FIGURE 35

Phialides and spores of various species of *Aspergillus*, reproduced @85%.

FIGURE 36

Spore masses of *Metarrhizium anisopliae* on agar. ×15

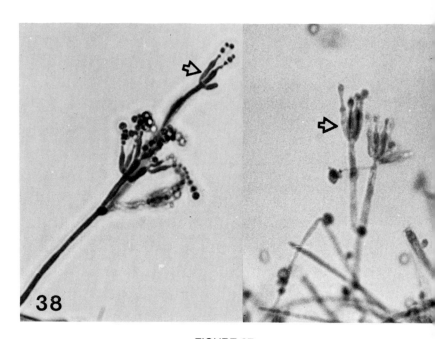

FIGURE 37

Conidia (elongate) (C) and conidiophores (arrow) of *Metarrhizium anisopliae.*
×240

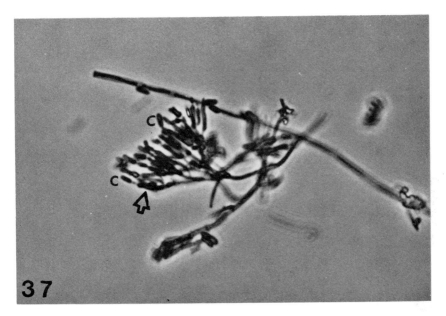

FIGURE 38

Phialides (arrows) and spores of a *Pencillium* sp. ×800, reproduced @ 90%.

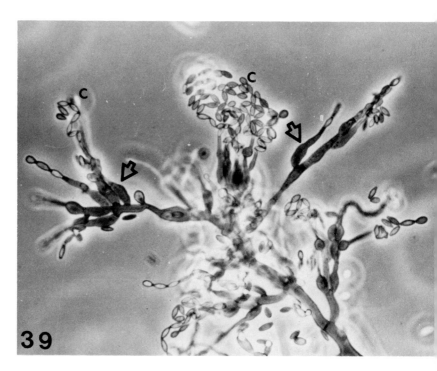

FIGURE 39

Conidia (C) and phialides (arrows) of *Paecilomyces farinosa.* ×800

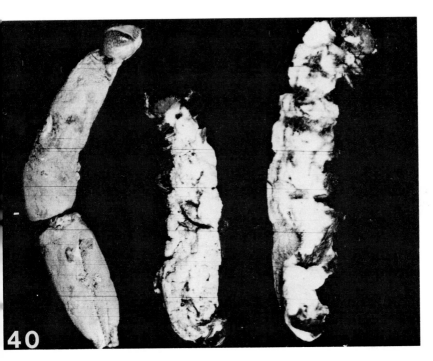

FIGURE 40

Nomuraea rileyi on the surface of diseased lepidopterous larvae. ×4

FIGURE 41

Phialides (arrows) and spores (C) of *Nomuraea rileyi.* ×760

BACTERIA

INTRODUCTION

The ubiquitous nature of bacteria make them the most abundant type of microorganism associated with insects. Thus, saprophytic, symbiotic, or pathogenic species of bacteria may be associated externally or internally with insects.

An example of the magnitude of these associations is shown by the relationships just between flies and bacteria (Greenberg, 1971). Then there are special associations, such as the one which results in the mortality of waterfowl that ingest fly maggots containing *Clostridium botulinum* (Hunter, 1970).

In this manual, we are concerned only with two groups of bacteria capable of causing disease. The first consists of the "true" pathogens which cause infection whenever they are ingested. The second group includes the potential pathogens which are omnipresent but cause infection usually when the insect is weakened or under stress.

TAXONOMIC STATUS

Previously placed in the class Schizomycetes, the bacteria have now been made a division in the kingdom Procaryotae, along with the blue-green algae (Buchanan and Gibbons, 1974). The main characteristics of this kingdom are the lack of a membrane separating the nucleoplasm from the cytoplasm, the lack of unit-membrane-bounded cytoplasmic organelles, and the presence of peptidoglycan in the cell walls.

Bacteria pathogenic to insects are all single cell organisms that reproduce primarily by fission. They may or may not be motile or form spores and most can be grown on artificial media.

LIFE CYCLE

In most infections, bacteria invade the host's hemocoel through the alimentary tract. This is certainly true for all entomogenous species of the genus *Bacillus,* in which spores serve as the infectious agents. For the potential pathogens, any break in the body wall or alimentary tract will serve as a port of entry.

Bacillus spores germinate and vegetative cells multiply first in the host's intestine before passing into the hemocoel. In most bacterial infections, once the vegetative cells reach the hemocoel, they produce a septicemia which invariably results in mortality. Depending on the bacterium present, spores or vegetative cells are formed in the cadaver and released into the environment.

Bacillus thuringiensis contains parasporal crystals which paralyze the gut of many lepidopterous larvae, thus allowing easy penetration by the vegetative cells. The known insect pathogens in the genus *Clostridium* cause disease by multiplying only in the insect gut and never enter the host's hemocoel.

Bacterial cells also may be introduced into the insect hemocoel by parasites or predators. The case of *Xenorhabdus nematophilus* is an excellent example. This bacterium occurs in the body of insects attacked by certain neoaplectanid nematodes and in the gut of the infective stage nematodes. The bacterial cells have no invasive power of their own and are introduced into the insect's hemocoel by infective stage neoaplectanid nematodes (Poinar and Thomas, 1967). Other genera and species of nematodes are also associated with related species of bacteria and have a similar relationship with insects.

CHARACTERISTICS OF INFECTED INSECTS

Since the gut is usually the initial organ affected in bacterial infections, the first signs of disease are related to feeding and assimilation. Loss of appetite, cessation of feeding, diarrhea, gut paralysis, and regurgitation are characteristic initial stages in many bacterial infections. Later the insect may appear sluggish (rarely irritable), have convulsions, and become uncoordinated; a general paralysis may set in, accompanied by septicemia and death. In a few instances, infected insects may show behavioral changes by moving to an elevated position or seeking refuge under leaves, etc.

Certain bacteria impart a characteristic color to the cadaver. For example, a red color suggests the presence of *Serratia marcescens*. Bee larvae infected with *Bacillus alvei* become yellow or gray, while those containing *B. larvae* become dark brown. The posterior portion of Japanese beetle grubs infected with *B. popillae* turns white. Most other bacterial infections turn the host brown-black, which is a color associated with bacterial decomposition.

FACTORS AFFECTING BACTERIAL INFECTIONS

With the obligate bacterial pathogens, certain conditions are necessary for successful infections. The so-called "milky disease" *Bacillus* species are infective only to members of the family Scarabaeidae. However, *B. thuringiensis* is known to infect insects in at least 4 orders, although lepidopterous larvae with a gut pH ranging from 9.0 to 10.5 are most susceptible.

Abiotic factors may also regulate infections. It has been known for some time that ultraviolet light is deleterious to spores and crystals of *B. thuringiensis*. Temperature, humidity, and formulation (when used as a biological insecticide) also affect the stability of *B. thuringiensis*. Environmental factors are especially important in governing infection by potential pathogens. A drop in temperature, starvation, stress resulting from crowding—all of these can make an insect more susceptible to pathogens that need the right "opportunity" to initiate infection.

METHODS OF EXAMINATION

A simple method of determining whether an insect has a bacterial infection is to examine a drop of its hemolymph under the microscope. Although bacterial rods and spores can be seen with bright field, most stages are more distinct with phase contrast.

Infections with bacterial pathogens are best diagnosed in the early stages of infection (before the insect has started to decompose), since then the majority of bacteria present in the blood or tissues will be those of the pathogen. The presence of bacteria of various sizes and shapes indicates the presence of saprophytic forms, and such cases are usually the most difficult to diagnose. It could be an infection by one of the potential pathogens, or the terminal stage of a virus or fungal infection. If a bacteriosis is suspected, then all bacteria present should be isolated and cultured. After pure cultures are obtained, identifications and pathogenicity tests can be conducted.

ISOLATION AND CULTIVATION

The first step in isolating bacteria from a diseased insect is to externally disinfect or "sterilize" the specimen to remove contaminating saprophytic forms. The anterior and posterior orifices of the specimen can be ligatured to keep the sterilizing solutions from entering the intestine and penetrating into the hemocoel. The following method has been used with satisfactory results. The specimen first should be dipped into 70% or 95% ethanol (wetting agent) for 2 sec. It can then be transferred to a solution of sodium hypochlorite (household bleach [5.25%] is suitable) for 3–5 min, then placed for an equal period in 10% sodium thiosulfate to remove the free chorine.

After the external surface has been sterilized, the specimen is rinsed in three changes of sterile distilled water and placed on a sterile dissecting dish. The specimen is then opened with sterile scissors or a scalpel by cutting the integument along a longitudinal dorsal or lateral line, with care being taken not to cut into the gut epithelium (all instruments should be sterilized, either by autoclaving, or by periodically dipping them into 70% ETOH and flaming off the alcohol). Blood and body fluids may be sampled with a sterile capillary tube, diluted in 2 ml sterile HOH or Ringer's solution, and then placed on NA, BHIA, or other suitable bacterial media by the streak plate method (see chapter on Techniques). Tissue may be examined by removing a small sample, placing it in 2 ml of sterile HOH or Ringer's and then triturating it with a sterile glass rod. The suspension is then streaked on a plate of nutrient or brain–heart infusion agar. The investigator may also want to inoculate AC medium if the presence of anaerobic pathogens is suspected. The agar plates may be inverted and incubated at room temperature or 30°C overnight, and then examined for well-isolated colonies. These colonies should represent pure bacterial strains and can be described on the basis of the following characteristics.

1. *Form* (of colony)
   a. Punctiform—under 1 mm in diameter, but visible to the naked eye.
   b. Circular—over 1 mm in diameter, round with smooth edges.
   c. Filamentous—growth consisting of interwoven or irregularly placed threads.
   d. Rhizoid—growth spreading out in an irregularly branched or rootlike manner.
   e. Spindle—usually subsurface colonies, larger in the middle than at the ends.
   f. Irregular—periphery is variable and nonuniform.
2. *Elevation*
   a. Flat—colonies are thin and raised little above agar surface.
   b. Raised—colonies are thick and noticeably raised above the agar surface.
   c. Convex—colonies are curved.
   d. Pulvinate—cushion-shaped colonies.
   e. Umbonate—colonies have a raised center; knoblike.
3. *Surface*
   a. Smooth—surface is even.

    b. Contoured—surface is irregular and smoothly undulating, similar to a relief map.
    c. Radiately ridged—ridges extend out from the center of the colony.
    d. Concentrically ringed—surface is marked with rings, one inside the other.
    e. Rugose—surface is wrinkled in appearance.
4. *Margin*
    a. Entire—a smooth margin.
    b. Undulate—wavy.
    c. Lobate—with rounded projections.
    d. Erose—irregularly notched.
    e. Filamentous—margin with long interwoven threads.
    f. Curled—composed of parallel chains of threads in wavy strands.
5. *Density*
    a. Opaque—light does not pass through.
    b. Translucent—light passes through, but is diffused so that objects beyond the colony cannot be discerned.
6. *Chromogenesis*—Refers to the production of pigment, such as the greenish pigment of some strains of *Pseudomonas aeruginosa,* and the red and orange red pigments of *Serratia marcescens.*

IDENTIFICATION

Practical methods of bacterial identification include both descriptive morphology and biochemical tests for the detection of metabolic products, enzymes, etc. Highly specialized techniques such as genetic analysis, serology, bacteriophage typing, and esterase patterns are beyond the scope of most insect pathology laboratories and are not covered here.

Because of time and material limitations, certain shortcuts for identification can be followed. Thus, a Gram-positive, motile, spore-forming rod isolated from the hemocoel of a diseased insect is almost certainly a *Bacillus*. Likewise, a polar-flagellate, oxidative, Gram-negative rod which produces a green fluorescent pigment is probably a *Pseudomonas* species and requires little further testing for genus.

Morphological examination is very important. Phase contrast microscopy provides a rapid method for detecting cell shape as well as motility. The Gram reaction is probably the most important staining procedure in bacteriology (see chapter on Techniques) and should be used routinely. Tests for flagellation and acid-fastness are also very useful.

Special diagnostic media may be very useful. On Tergitol-7 medium with TTC (triphenyltetrazolium chloride), *Escherichia coli* forms yellow colonies against a yellow background, while *Pseudomonas aeruginosa* and other Enterobacteriacae form red colonies against a blue background. A diagnostic medium is available for *Streptococcus faecalis* (Meade, 1963), while *P. aeruginosa* can be determined on regular agar plates with the cytochrome oxidase test (Schaefer, 1961).

There are several species of *Bacillus* and *Clostridium* which do not grow well or at all on artificial media. Thus, careful comparison should be made with the bacteria found in the insect and the isolates growing on culture media to be certain the latter are not contaminants.

TESTING FOR PATHOGENICITY

Experimental infections are often necessary to establish the pathogenicity of an unknown isolate. These can best be done by introducing the bacteria in question into the gut of test insects. A glass-tipped syringe can be used for forced feeding or the inoculum can be mixed with food. Koch's rules of pathogenicity can be followed if the results are successful.

STORAGE

From the practical standpoint, the most important entomogenous bacteria belong to the genus *Bacillus*. The spores of most of these species are resistant and survive well if stored in a cool, dry location. Non-spore-forming bacteria can be maintained on simple bacteriological media or lyophilized.

LITERATURE

The eighth edition of *Bergey's Manual* (Buchanan and Gibbons, 1974) contains the latest views on bacterial classification and includes keys to the genera of bacteria.

For entomogenous bacteria, Steinhaus (1949) provides a good general introduction to the topic and more specific coverage is provided by Bucher (1963), Dutky (1963), Heimpel and Angus (1963), and Lysenko (1963). The use of bacteria for insect control was discussed by Falcon (1971) while other subjects on the toxins and host spectrum of *B. thuringiensis* were included in Burges and Hussey's *Microbial Control of Insects and Mites* (1971). Faust (1974) provided a thorough review of the bacterial diseases of insects and Afrikian (1973) recently published a book on entomogenous bacteria.

KEY TO COMMON GENERA AND SPECIES

1. Vegetative cells coccoid in shape (Figure 42), young cultures Gram positive, spores absent—**2**
1. Vegetative cells rōd shaped (may be very short), Gram negative, positive or variable; spores present or absent—**3**
2. Cells often forming chains in nutrient broth; catalase negative—*Streptococcus* Rosenbach (Figure 42). *S. faecalis* Andrews and Horder is a commonly encountered coccus causing insect disease. See Doane and Redys (1970) for a discussion of this species. *S. pluton* (White) is associated with European foulbrood of honeybees. See Bucher (1963) for a discussion of this genus as related to insect disease.
2. Cells occurring singly, in pairs, tetrads, irregular clusters, or cubical packets; catalase positive—*Staphylococcus* Rosenbach and *Micrococcus* Cohn. Members of these genera may occur in diseased insects, but have never been implicated as causal agents.

3. Spores present; young cultures Gram positive (*Bacillus sphaericus* is gram variable)—**4**

3. Spores absent; young cultures Gram negative—**10**

4. Crystalline parasporal body present (may be absent in some strains) (Figure 43)—**5**

4. Crystalline parasporal body absent (Figure 44)—**6**

5. Good growth on artificial media under aerobic conditions; catalase positive; found mainly in larvae of Lepidoptera—*Bacillus thuringiensis* Berliner (Figure 43). See de Barjac and Bonnefoi (1973) and Heimpel (1967) for a discussion of this group.

5. Poor growth on artificial media under aerobic conditions; catalase negative; found in Scarabaeidae—*Bacillus popillae* Dutky (Figure 45). This species contains several varieties, one of which lacks the parasporal body (Wyss, 1974). See Dutky (1963) for a discussion of this species and Wille (1956) for a discussion of *B. fribourgensis*.

6. Little or no growth on artificial media under aerobic conditions; (obligate or facultative anaerobes); bacteria restricted to gut of diseased insect; thus far natural infections found in *Malacosoma* sp. (Lepidoptera)—*Clostridium* Prozmowski (Figure 46). See Bucher (1961) for a discussion of two species in this genus that cause insect diseases.

6. Good growth on artificial media under aerobic conditions; bacteria occur in hemocoel of diseased insects; found in a variety of insects—**7**

7. Sporangia definitely swollen (Figures 47, 48); spores oval or spherical—**8**

7. Sporangia not swollen; spores ellipsoidal or cylindrical—**9**

8. Spores nearly spherical; pathogenic for mosquito larvae—*Bacillus sphaericus* Neide (Figure 47). See Kellen *et al.* (1965) for a discussion of this species.

8. Spores oval; cause of American foulbrood in honeybees—*Bacillus larvae* White (Figure 48). See Heimpel and Angus (1963), Cantwell (1974), and Glinski (1968) for a discussion of this species.

9. Acetylmethylcarbinol not produced (Voges Proskauer [VP] test is negative); thought to cause disease in silkworm and the scarabeid, *Melolontha melolontha—Bacillus megaterium* de Bary. See Heimpel and Angus (1963) for an account of this species.

9. Acetylmethylcarbinol produced (VP test is positive)—*Bacillus cereus* Frankland and Frankland (Figure 44). Saprophytic species of *Bacillus* will also key out here, but only *B. cereus* has been associated with insect disease. See Heimpel and Angus (1963) for a discussion of this species.

10. Cells polar flagellate; colonies on agar give a rapid positive reaction to the cytochrome oxidase test (see chapter on Techniques)—*Pseudomonas aeruginosa* (Schroeter) (Figure 49). See Bucher (1963) for an account of the pseudomonads associated with insects.

10. Cells peritrichously flagellate or without flagella; cytochrome oxidase test negative—**11**

11. Bacteria associated with the presence of rhabditoid nematodes in the hemocoel; large rods (2–10 μm long)—*Xenorhabdus nematophilus* Poinar and Thomas (Figure 50). This bacterium is associated with neoaplectanid nematodes and its nature with insects is discussed by Poinar and Thomas (1967). A closely related strain (or species) is associated with nematodes of the genus *Heterorhabditis* (= *Chromonema*).

11. Bacteria not associated specifically with rhabditoid nematodes; small rods (0.5–2.0 μm long)—**12**

12. Often producing various shades of orange to red pigments in the insect or on agar plates; flagella close coiled when stained by Leifson's method (see chapter on Techniques); does not ferment arabinose—*Serratia marcescens* Bizio (Figure 51). See Steinhaus (1959) and Bucher (1963) for associations of this species with insects.

12. Nonchromogenic; flagella not close coiled; may or may not ferment arabinose—**13**

13. Phenylalanine deaminase negative; VP usually positive—*Enterobacter* Hormaeche and Edwards. Species in this genus have been reported as potential pathogens of grasshoppers; however, their invasiveness under natural conditions is ques-

tionable. See Bucher (1959) and Faust (1974) for a discussion of these forms.

13. Phenylalanine deaminase positive; VP usually negative— *Proteus* Hauser. Species in this genus have been reported as potential pathogens of grasshoppers; however, their invasiveness under natural conditions is questionable. See Faust (1974) for a discussion of these forms.

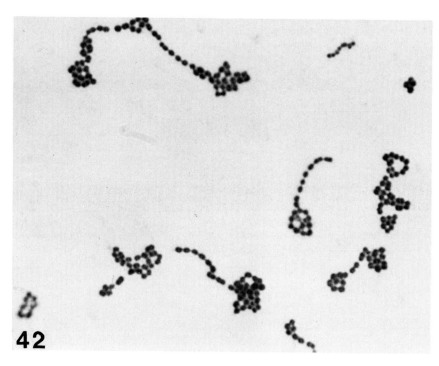

FIGURE 42

Coccoid vegetative cells of *Streptococcus* sp. (methyl blue stain). ×1800

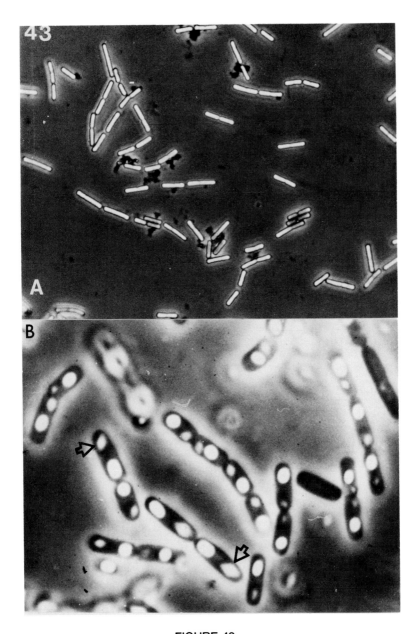

FIGURE 43

Vegetative cells (A) (×1600) and sporangia (B) (×2560) of *Bacillus thurin-giensis.* Note crystalline parasporal body (arrows), reproduced @ 90%.

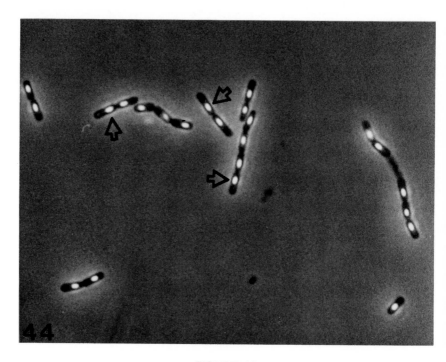

FIGURE 44

Nonswollen sporangia of *Bacillus cereus*. Note spores (arrows). ×1600

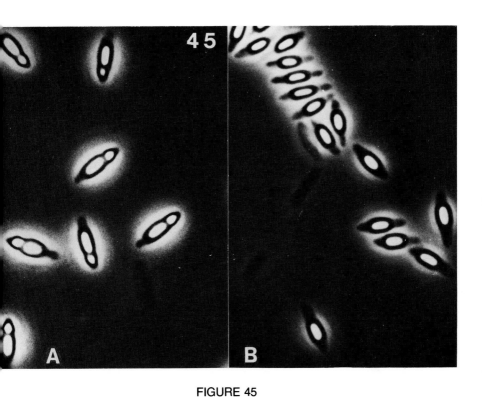

FIGURE 45

Sporangia of *Bacillus popillae,* (A) with parasporal body, (B) without parasporal body (courtesy of C. M. Splittstoesser). ×3400, reproduced @ 90%.

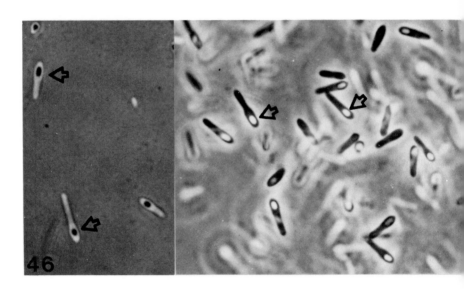

FIGURE 46

Sporangia of *Clostridium sporogenes* with terminal spore (arrows). ×2100, reproduced @ 90%.

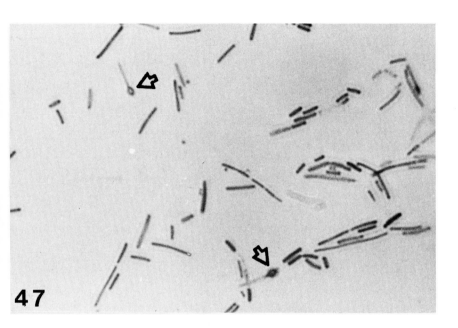

FIGURE 47

Vegetative cells and swollen sporangia of *Bacillus sphaericus.* Note nearly spherical spore (arrows).   ×1024

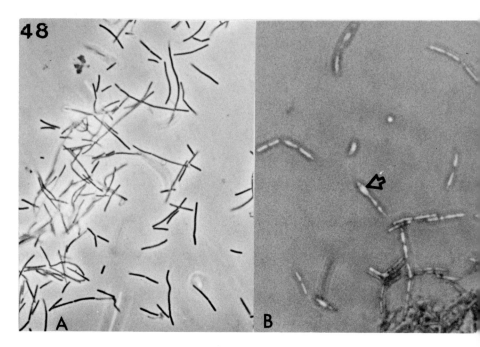

FIGURE 48

Vegetative cells (A) (×760), and swollen sporangia (B) (×1540) containing oval spores (arrow) of *Bacillus larvae,* reproduced @ 90%.

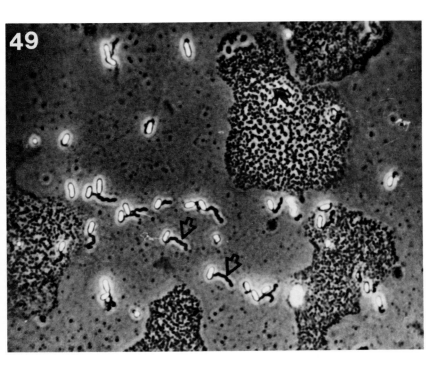

FIGURE 49

Vegetative cells of *Pseudomonas aeruginosa* with polar flagella (arrows).   ×1800

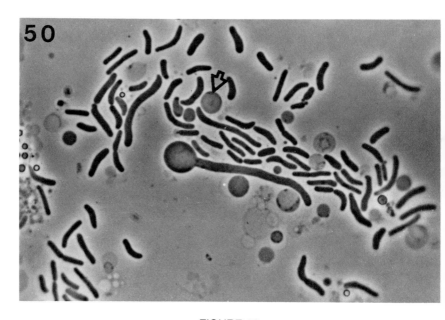

FIGURE 50

Variable sized vegetative cells and associated protoplasts (arrow) of
Xenorhabdus nematophilus. ×1500

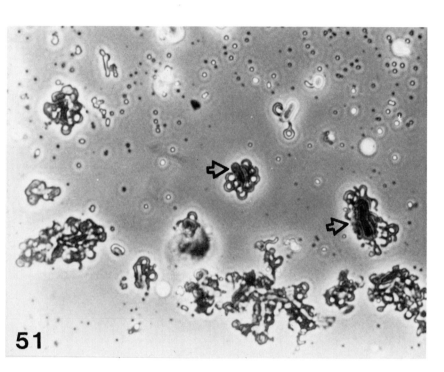

FIGURE 51

Vegetative cells (arrows) of *Serratia marcescens* showing close-coiled
peritrichous flagella.   ×1800

VIRUSES

INTRODUCTION

Insect virology is a very dynamic field due to the continuous discovery of new viruses and new host–pathogen relationships. While viruses originally were thought to be restricted to a few specific insect groups, it is now apparent that they can infect representatives of many insect orders. Viruses very similar to those attacking insects have also been reported in mites, shrimp, and even some vertebrates. Widespread use of these viruses is now being discussed, and safety considerations for the baculoviruses have already been examined (Summers *et al.*, 1975).

Viruses which cause diseases in plants and vertebrates may be transmitted by insects. These vectors may carry the virus mechanically or actually serve as a second host. In the latter case, the virus multiplies in the vector and may act as a pathogen by reducing longevity, damaging tissues, etc. Thus, Mims *et al.* (1966) discovered that salivary glands of *Aedes aegypti* were destroyed by an arbovirus. Likewise, plant viruses may produce cytopathic changes, a shortened life span, reduction in reproductive ability, or even death in their insect vectors (Maramorosch, 1968a). However, most of the viruses which are vectored by insects are not pathogenic to their host and will not be discussed further in this manual.

TAXONOMIC STATUS

Viruses can be defined as submicroscopic, obligate, intracellular, pathogenic entities. Experts claim that about 450 different viruses have been isolated from insects and mites and that about 90% of these have inclusion bodies (= occluded) (Ignoffo, 1974). It is beyond the scope of this manual to treat all of the above, especially since many of these have yet to be characterized and classified.

Insect viruses were originally classified on the morphology of their inclusion bodies and virions or virus particles. Host group and tissue affinity were also characteristic features. At one stage, virus groups were given generic names such as *Borrelinavirus, Smithiavirus,* and *Bergoldiavirus,* which corresponded to the nuclear polyhedrosis (NPV), cytoplasmic polyhedrosis (CPV), and granulosis viruses (GV), respectively. However, as more basic biochemical and structural data were obtained on insect and other viruses, it became possible to compare all known viruses and a new set of "genera" were proposed. Thus, the nuclear polyhedrosis and granulosis viruses were placed together in the "genus" *Baculovirus.* This classification, which is followed here, is discussed by David (1975).

From the practical standpoint of identification, insect viruses fall into two groups, those possessing inclusion bodies (usually just visible in the light microscope) and those without inclusion bodies (visible only with the electron microscope).

Those without inclusion bodies contain virus particles (virions) which may be rod, bullet, oval, or isometric shaped. When inclusion bodies occur, the virions usually are embedded within them.

LIFE CYCLE

Except for specific cases of transovarial transmission, insect viruses generally enter a host through the mouth and digestive tract. The virus particles which are ingested directly or released from inclusion bodies infect or pass through the gut epithelium to enter susceptible host tissues. Some viruses appear to infect specific tissues, while others are capable of infecting most, if not all, tissues of the host. While most viruses produce acute infections, others are occult and produce an inapparent disease which cannot be detected by external examination of the host. Occult viruses may be induced to the acute pathogenic stage by subjecting the insects to stress conditions such as crowding or starvation. By activating latent infections, these conditions are probably responsible for the sudden and dramatic outbursts of virus diseases. The infection process of various insect viruses has been summarized by David (1975), Harrap (1973), Delgarno and Davey (1973), and Vaughn (1974).

CHARACTERISTICS OF INFECTED INSECTS

Insects suffering from virus infections may exhibit morphological, physiological, and behavioral symptoms. The extent and type of symptoms depends on the virus and host involved. Lepidopteran larvae infected with nuclear polyhedrosis viruses may show behavioral abnormalities such as moving toward the tops of plants, where they cease feeding and become flaccid. Death quickly follows with disintegration of internal tissues and release of inclusion bodies which may cloud the hemolymph. Pupae may show similar symptoms when late larval instars become infected.

Sawfly larvae infected with NPV may exhibit a faint, yellow discoloration (especially on the third to fifth abdominal segments), lose their appetite, and become inactive. A brown or milky fluid is often exuded from the anus.

Larvae of *Tipula paludosa* (Meigen) infected with NPV become lighter as the disease progresses and finally turn chalky white.

Lepidopteran larvae infected with cytoplasmic polyhedrosis viruses usually are retarded in growth. As the disease progresses, the infected midgut may be visible through the integument as a pale yellow or whitish area. In later stages, polyhedra are often regurgitated or passed out with the feces. The skin of diseased larvae usually does not rupture as with nuclear polyhedrosis infections.

Symptoms associated with granulosis viruses are nonspecific and vary considerably from one insect to another. The first symptom is usually a paling in color, followed by loss of appetite. There may be a mottling of the integument, and very often the ventral surface will become progressively pale whitish or milky yellow due to the infected fat bodies. As infected tissues disintegrate, large numbers of capsules are released, and the hemolymph becomes turbid and milky. In some species, there may be a liquification of internal tissues after death, and when the epidermis is infected, the integument becomes very fragile, similar to a nuclear polyhedrosis. However, if the epidermis is not affected, the integument remains relatively firm.

Larvae infected with iridoviruses can often be recognized by an opalescent, iridescent, blue, green, or brown color of the infected tissues.

The symptoms of other nonoccluded viruses seem to be peculiar to their host and the type of disease produced. For example, the inherited "sigma virus" of *Drosophila* sp. leaves the flies sensitive to carbon dioxide. Honey bees infected with chronic and acute paralysis viruses exhibit symptoms of trembling and loss of coordination. Honeybee larvae which die from the sacbrood virus are extended lengthwise along the floor of the cell with the head darker than the rest of the body.

METHODS OF EXAMINATION

If a virus disease is suspected, an examination of the host's tissues should be made with the light microscope. Since the majority of reported insect viruses belong to the occluded type, their variously shaped inclusion bodies are usually visible under the light microscope. The inclusion bodies of nuclear and cytoplasmic polyhedrosis viruses appear refringent (shining white) under bright field and phase contrast, whereas the capsules of granulosis viruses appear white in bright field and gray in phase contrast. Small inclusion bodies may exhibit Brownian movement in wet mounts.

Uric acid crystals (Figure 2) often occur in diseased insects and may resemble inclusion bodies. However, urate crystals are birefringent and often give a characteristic cross appearance under polarized light, while inclusion bodies are monorefringent. The latter can be distinguished from spherical fat droplets (Figure 1) which turn red in the presence of aqueous Sudan III (10–15 min). The inclusion bodies do not stain with Sudan; however, their ability to dissolve in strong alkali (1 N NaOH) is characteristic (Figure 58). Inclusion bodies can be further demonstrated in diseased tissues by special staining methods, such as the Feulgen–Schiff reaction, slow Giemsa staining with acid hydrolysis, and the iron hematoxylin methods described by Huger (1961) (see chapter on Techniques). The inclusion bodies (spheroids) of pox viruses often swell and darken when stained with lactophenol–cotton-blue. Sikorowski *et al.* (1971) described a method for detecting polyhedra of CPV in *Heliothis*. These methods are described in the chapter on Techniques.

Noninclusion viruses can only be detected with the electron microscope. However, noninclusion virus infections sometimes can be determined from various signs and symptoms of infected hosts, along with the absence of other pathogens.

ISOLATION

There are several methods of isolating inclusion viruses for further studies. The simplest of these involves placing the diseased insects in a culture tube with water. After two or three days, the inclusion bodies will accumulate as a white layer on the bottom of the tube. Cell and tissue remnants, bacterial cells, and other breakdown products can be separated from the inclusion bodies by repeated washing and differential centrifugation. If the virus is to be used for infection studies, the pellet can be treated with antibiotics for 24 hr and washed several times with sterile distilled water, although differential centrifugation generally removes most bacteria. Inocula used for injection studies should be assayed for viable bacteria or fungi by culturing an aliquot on an enriched agar medium such as brain–heart infusion agar or AC medium (see chapter on Techniques). A variation which is particularly helpful in purifying very small inclusion bodies, such as granulosis capsules and small cytoplasmic polyhedra, is to mix the triturated diseased insect with an equal volume of carbon tetrachloride, shake vigorously, and centrifuge at 3000 rpm for about one hour. A layered plug of inclusion bodies forms between the CCL4 (bottom) and the water phase, while fat accumulates on top of the water phase. The water and fat can be decanted off without disturbing the plug, and the inclusion bodies, which form the top layer of the plug, can be carefully washed or scraped off with a small spatula and suspended in water. Further purification may be accomplished by repeated washing and differential centrifugation. If convenient, a sucrose-gradient centrifugation will give a purer preparation of capsules and polyhedra than the CCL4 method.

With the nonoccluded iridoviruses, the first step in the isolation process is to dissect out the infected tissues (identified by their iridescent color) triturate them in buffer and leave them for 24 hr. The action of autolytic enzymes will further disintegrate the tissues and free the virus particles. The suspension should then be centrifuged at 3000 rpm for about 30 min, and the virus-containing supernatant saved. The supernatant can be freed of most bacteria by differential and sucrose-gradient centrifugation or passage through a 0.45-μm millipore filter. Any fat will settle on top of the supernatant and must be removed before filtration. Ultracentrifugation of the above supernatant at 10,000 rpm will concentrate the virus in a pellet which can then be washed and purified by differential centrifugation.

Another method of obtaining almost pure iridescent virus particles is to infect lepidopteran larvae containing large silk glands (e.g., *Bombyx mori* L., *Pseudaletia* sp., or *Galleria mellonella* L.). After the infection is well advanced, the heavily infected blue silk glands can be dissected out and the virus particles removed.

Noniridescent, noninclusion viruses are much more difficult to detect and purify. A general method which can be followed is that used by Bailey *et al.* (1964) for the isolation of adult bee paralysis and sacbrood viruses of honeybees. Whole insects, or washed parts of them, are ground in tap water and a quarter volume of CCL4. The resulting emulsion is coarsely filtered (e.g., through cheesecloth) and the filtrate purified by centrifugation at 8000g for 10 min. The water phase containing the virus is at the top and can be used as is or purified and concentrated by high-speed centrifugation.

IDENTIFICATION

Most inclusion viruses can still be identified to group on the basis of the morphology of the inclusion body and enclosed virus particles. However, size and shape are no longer sufficient characters for all of the nonoccluded viruses, and the type and strandedness of the nucleic acid plus other biochemical characteristics are often necessary.

Insect viruses have been placed in six "genera." The cytoplasmic polyhedrosis viruses are still to be assigned to a "genus." The classification of insect viruses will certainly change in the future; however, a wise move has been made to treat the invertebrate viruses in the same general classification as the vertebrate and plant viruses (Wildy, 1971).

In identifying insect viruses, characteristics of the disease are also important. The type of tissue infected, as well as any abnormal symptoms and the hosts involved are all useful aids. Aside from specialized instances, however, host groups cannot be used for virus identification. There are now too many instances of similar viruses appearing in unrelated hosts.

TESTING FOR PATHOGENICITY

Since viruses normally enter their insect hosts per os, the simplest method for obtaining experimental infections is to introduce the virus particles into the mouth of a test insect. A calibrated hypodermic syringe can be used for introducing a measured amount of virus suspension into the mouth of test insects. An indirect approach is to contaminate the insect's food; for example, foliage can be immersed in virus suspensions and fed to insect larvae. For intrahemocoelic injections (for nonoccluded viruses), the virus suspension must be purified or freed of bacteria that could multiply and destroy any potential host. This can be done by first passing the suspension through a 0.45-μm millipore filter or treating it with a 5000 unit/ml mixture of penicillin–streptomycin.

STORAGE

Inclusion bodies can be stored either in a purified condition, within the host tissues under refrigeration (5°C) or frozen. All insect viruses apparently can survive lyophilization (freeze drying), and this may prove to be the most efficient method of storage.

LITERATURE

Basic coverage of insect viruses is presented by Smith (1967), who later reviewed the polyhedroses and granuloses of insects (Smith, 1971). Aruga and Tanada (1971) edited a book on the cytoplasmic polyhedrosis virus of the silkworm and Vaughn (1974) presented a general review of the virus diseases of insects. Other groups of insect viruses are discussed in works edited by Steinhaus (1963) and Maramorosch (1968a,b), while Bergoin and Dales (1971) presented a detailed account of pox viruses of invertebrates and vertebrates.

Other discussions of insect viruses include the replication of baculoviruses, cytoplasmic polyhedrosis viruses, and iridoviruses by Delgarno and Davey (1973), a general discussion of virus infection in invertebrates by Harrap (1973), and associations between viruses and Diptera (Marshall, 1973), viruses and Lepidoptera (Longworth, 1973), viruses and Hymenoptera (Bailey, 1973b), and viruses and leafhoppers (Sinka, 1973).

Recent advances in insect virology with current views on classification are presented by David (1975). Discussion on the use of viruses for insect control have been presented by Stairs (1971), Ignoffo (1968, 1974), and Bailey (1973a). Martignoni and Iwai (1975) have compiled a useful computer-based catalog of viral diseases of insects and mites. Summers *et al.* (1975) discuss the safety considerations of baculoviruses in insect control.

KEY TO GROUPS OF VIRUSES

1. Inclusion bodies (protein crystals enclosing virus particles) present (visible with the light microscope) (Figures 52, 54, 56, 58)—2

1. Inclusion bodies absent; only virions or virus particles formed in host cells (visible only with the electron microscope) (Figures 60, 61)—**5**
2. Inclusion bodies (capsules) ovoid, ellipsoidal, small; 0.2–0.5 μm in length; appear dark under phase contrast; thus far found only in larvae of Lepidoptera—Granulosis viruses (Baculoviruses) (Figures 52, 53)
2. Inclusion bodies irregular, polyhedral (0.2–20.0 μm in diameter); appear light under phase contrast; occur in representatives of many orders—**3**
3. Inclusion bodies (spheroids) ovoid, 2.0–20.0 μm in diameter; noninfective spindle-shaped bodies may also be present; virions ovoid–cuboid shaped, contain DNA—Poxviruses (Figures 54, 55).
3. Inclusion bodies irregular, polyhedral; 0.2–15.0 μm in length, virions rod-shaped or isometric, contain DNA or RNA—**4**
4. Inclusion bodies usually 0.2–2.5 μm in diameter, formed in the cytoplasm of midgut epithelial cells; virions isometric, contain RNA—cytoplasmic polyhedrosis viruses (Figures 56, 57).
4. Inclusion bodies usually 0.5–15.0 μm in diameter, formed in the nuclei of various tissues; virions rod-shaped, contain DNA—nuclear polyhedrosis viruses (Baculoviruses) (Figures 58, 59).
5. Diseased insects and/or infected tissues showing iridescence with reflected light; virions isometric, containing DNA, 130–240 nm in diameter—Iridescent viruses (Iridoviruses) (Figure 60).
5. Iridescence lacking, virions isometric or elongate—**6**
6. Virions bullet-shaped or bacilliform; 70 × 140 nm; symptom in adult *Drosophila* (only known host) is a sensitivity to CO_2; contain RNA—*Drosophila* sigma virus (a Rhabdovirus)
6. Virions isometric, found in a variety of insects (Figure 61)—**7**
7. Virions containing DNA—Parvoviruses (includes *Galleria* densonucleosis virus)
7. Virions containing RNA—Enteroviruses (includes viruses causing diseases in adult (paralysis) and larval (sacbrood) bees, as well as in representatives of Lepidoptera, Isoptera, and Orthoptera).

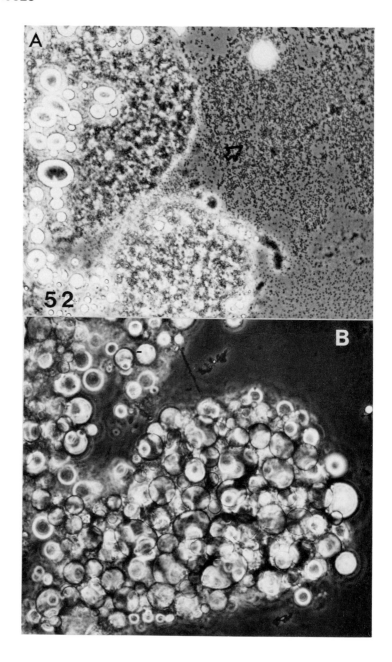

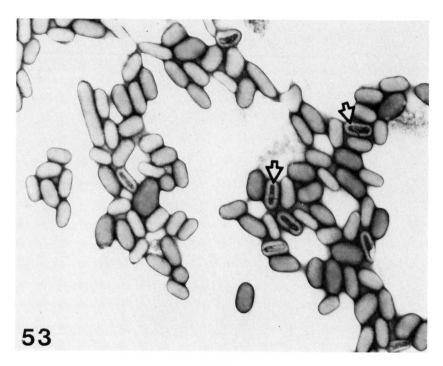

FIGURE 53

Electron micrograph of a granulosis virus; arrows show virus rods within capsules
(courtesy of Roberta Hess). ×27,400

FIGURE 52

(A) Fat body infected with a granulosis virus; note liberated capsules (arrow).
(B) Normal fat body. ×640, reproduced @ 95%.

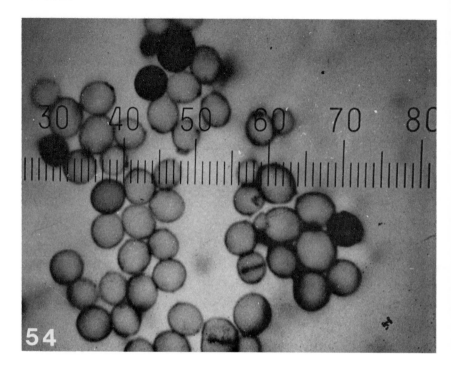

FIGURE 54

A pox virus (courtesy of R. Goodwin). ×2,000

FIGURE 55

Electron micrograph of a pox virus. Note enclosed virions (arrows). (A) ×17,000
(B) ×75,000 (courtesy of R. Goodwin).

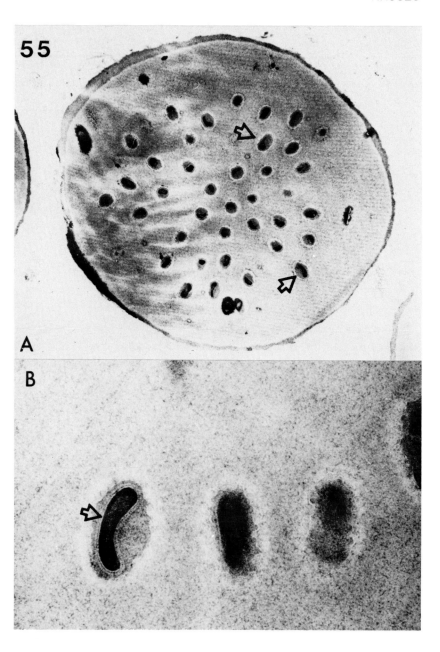

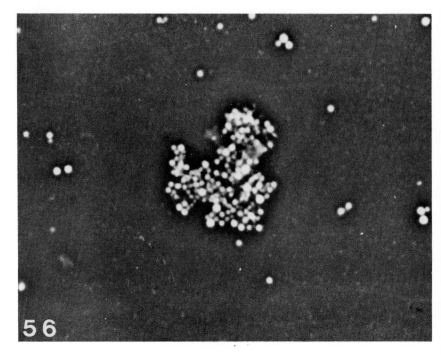

FIGURE 56

Cytoplasmic polyhedrosis virus. ×2,000

FIGURE 57

Electron micrographs of cytoplasmic polyhedrosis viruses. (A) Several polyhedra (×14,500). (B) Single polyhedron (×64,132). Arrows show virus particles (courtesy of Ruth Sluss), reproduced @ 80%.

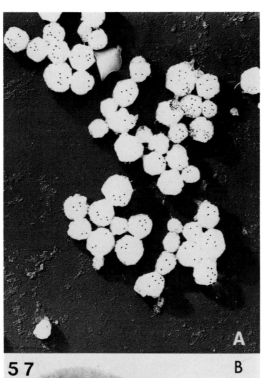

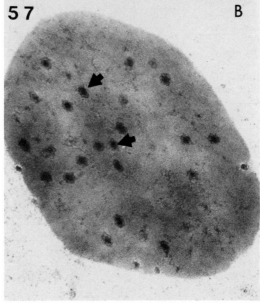

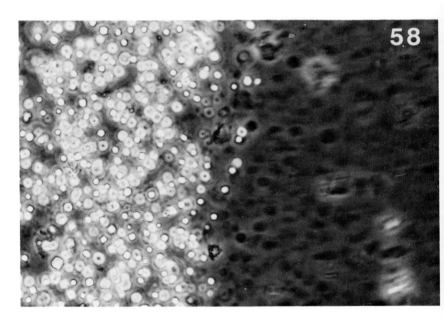

FIGURE 58

Polyhedra of a NPV being dissolved with NaOH. ×500

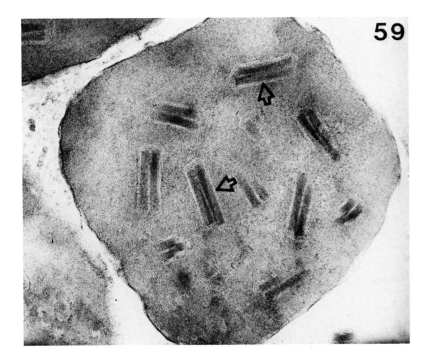

FIGURE 59

Electron micrograph of a polyhedron of a NPV. Arrows show virus particles
(courtesy of Roberta Hess). ×53,500

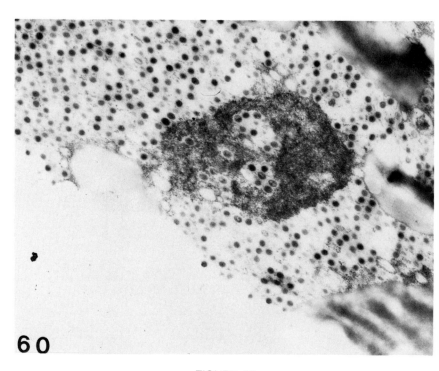

FIGURE 60

Electron micrograph of an iridescent virus in insect tissue (courtesy of Ruth Sluss).
×13,500

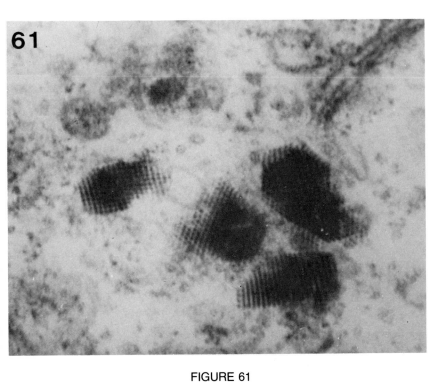

FIGURE 61

Electron micrograph of a paracrystalline array formed from adult honeybee paralysis virus particles (courtesy of Ruth Sluss). ×84,000

PROTOZOA

INTRODUCTION

Representatives of many groups of protozoa have associations with insects. These associations range from phoresis to obligate parasitism, often with host debilitation or mortality as the final result. Some symbiotic protozoa found in the gut of insects (e.g., termites, wood roaches) are important for the well-being of their bearer, and others, seemingly innocuous for their insect vectors, cause serious diseases in man and domestic animals.

Some groups of insect-pathogenic protozoa may also infect other invertebrates and even vertebrates. For that reason, a word of caution has been raised concerning the widespread use of some protozoa as biological control agents. However, other pathogens appear more specific and already have been used as biological control agents in several instances. Occasionally, protozoan pathogens may attack parasites which occur in the original host. Thus, aside from infecting its normal weevil host, *Anthonomus grandis,* the protozoan *Mattesia grandis* also attacks a braconid parasite of the boll weevil (McLaughlin, 1965). Veremtchuk and Issi (1970) found a *Nosema* which normally infects *Pieris brassicae* L. but will also attack neoaplectanid nematodes parasitizing the lepidopterous host.

The minute size of many protozoa has prevented their detailed study, and the life cycles and developing stages of relatively few are known. An urgent need in this area of study remains unfilled.

TAXONOMIC STATUS

The Protozoa are generally considered to have the status of phylum. However, some have placed these organisms in a separate kingdom, the Protoctista (or Protista) (Whittaker, 1969). The latter system of classification (multikingdom) has the advantage of including organisms that resemble plants as well as animals in a broad category since the standard definition of a protozoan as a "unicellular animal with all body functions being performed by that cell" leaves much to be desired.

The classification used here is based on the report by Honigberg *et al.* (1964) who split the phylum Protozoa into several subphyla and lower categories. The majority of insect pathogenic protozoa belong to the subphyla Sporozoa and Cnidospora. Other subphyla containing insect pathogens are the Sarcomastigophora and Ciliophora. Many of the higher levels of classification are undergoing revision and some authors propose the erection of more than one phylum for the Protozoa.

LIFE CYCLE

The infective stages (spores, cysts, etc.) of most insect pathogenic protozoa are ingested and pass into the host's alimentary tract (transovum and transovarial transmission may also occur). Entry into the hemocoel supposedly occurs through the midgut wall; however, very little is known about the initial steps of infection. Usually the first signs of infection are the presence of the developing stages in the gut epithelium, fat body, malpighian tubules, or hemolymph.

Reproduction may occur asexually by binary fission (flagellates and ciliates), by schizogony or multiple fission (coccidians) or by gamogony (gregarines). Sexual reproduction in the Sporozoa consists of the fusion of two gametes to form a zygote which in turn undergoes repeated divisions. In some ciliates and suctorians reproduction by conjugation, autogamy, endomixis, and cytogamy may occur.

CHARACTERISTICS OF INFECTED INSECTS

There are very few specific symptoms in insects suffering from protozoan infections. Most are general symptoms which also occur in insects infected with a variety of pathogens. These include small size, morphological deformities, lethargy, difficulty in molting, reduction of feeding, loss of balance, and production of a white fecal exudate. Hosts with a transparent cuticle often have white opaque or swollen areas on the cuticle, whereas other insects indicate a protozoan infection by the presence of black spots on their integument. However, in many cases, collapse or mortality is the first indication of a protozoan infection, which must be confirmed by direct microscopic examination. Burkholder and Dicke (1964) devised a method of detecting insects infected with *Mattesia dispora* with ultraviolet light. This method might be extended to other protozoan infections as well.

METHODS OF EXAMINATION

The simplest way to detect a protozoan infection is to examine a smear mount using fat body, malpighian tubules, gut epithelium, or hemolymph under phase contrast or bright field microscopy. The mature spores or spore-containing bodies are the easiest stages to distinguish. Staining with Giemsa or hematoxylin will show the developing stages more clearly (see chapter on Techniques). Both spores and developing stages are important for diagnosis.

It is sometimes difficult to differentiate protozoan spores from artifacts or resistant stages of other organisms. Most inclusion bodies of viruses dissolve in weak NaOH whereas protozoan spores do not. Many fungal spores resemble those of protozoa and, aside from observing germination in the former, there is sometimes no sure way of separating the two. Fungal spores may be colored and are generally more uniform in texture and content. Giemsa stain may demonstrate the fine internal structure of protozoan spores, and extrusion of the polar filament is a simple, quick method for confirming the presence of a microsporidan. The presence of a protozoan in a diseased insect does not imply it is the cause of the disease, however. It must be identified and compared with those forms known to cause insect disease or tested for pathogenicity. Thus, flagellates, amoebae, and ciliates may be associated with insects, but are not necessarily pathogenetic.

ISOLATION AND CULTIVATION

Protozoa can best be isolated by removing a sample of hemolymph or tissue from an infected insect. However, only a few pathogenic protozoa can be grown in artificial media (e.g., *Tetrahymena, Malpighamoeba, Plasmodium*) and these methods are not practical for the general diagnostician.

IDENTIFICATION

The minute size of many protozoans makes their examination difficult. However, attempts should be made to determine the presence and type of locomotor organelles, the size and structure of the spores or vegetative stages, the number of spores in a "cyst," the number of sporozoites in a spore and the presence or absence of a polar filament (diagnostic character for the Microsporida). Many protozoans appear to be host and tissue specific so it is important to identify the insect and tissue infected (fat body, gut epithelium, malpighian tubules, etc.). More protozoologists are using spore ultrastructure as diagnostic characters for new genera. Thus, revisions of existing systems of classification will unfortunately be based on characters visible only with the electron microscope.

TESTING FOR PATHOGENICITY

Since most entomogenous protozoa enter their hosts through the digestive tract, pathogenicity tests should be made by introducing protozoan spores into the mouth of the test insect. Diseased insects can be triturated in distilled water, and after removing the debris by straining and differential centrifugation, the spore suspension can be mixed with food or sprayed on unhatched eggs or plants.

It may be desirable to add a feeding stimulant. With large insects, a suspension of spores can be introduced directly into the mouth cavity with a calibrated syringe.

STORAGE

After purification, the resistant stages of some protozoans can be stored under refrigeration. For the more fastidious forms lacking resistant stages, continuous transfer from host to host may be the only method of maintenance.

LITERATURE

The text *Protozoology* by Kudo (1966) presents a general account of the Protozoa and covers most of the insect forms. The basic and advanced texts of Steinhaus (1949, 1963) also cover the insect-associated protozoa and W. M. Brooks (1974) recently presented a detailed review of protozoan infections in insects. A manual for the identification of protozoa was prepared by Jahn (1949) and practical aspects of the use of protozoa for insect control have been presented by McLaughlin (1971, 1973). More specific coverage of the microsporida has been presented by Kudo (1924) and Weiser (1961).

KEY TO COMMON GENERA

1. Active stages containing cilia, flagella, or pseudopodia; non-motile resting states not containing spores—**2**
1. Active stages absent; no stages containing cilia, flagella, or pseudopodia; resting stages represented by spores or cysts containing spores—**8**
2. Cilia and/or sucking tentacles present (Subphylum Ciliophora, Class Ciliatea) (Figure 62)—**3**
2. Cilia and sucking tentacles absent; motile stages containing flagella or showing amoeboid movement; resting stages may be composed of thick-walled cysts commonly found in the intestine or malpighian tubules of the host (Subphylum Sarcomastigophora)—**6**

3. Cilia present only in juveniles (usually not encountered), and absent in mature stages, which contain tentacles; usually sessile forms attached to the exoskeleton of aquatic insects by a non-contractile stalk—Subclass Suctora (Figure 63). Members of the genera *Discophrya* Lachmann, *Periacineta* Collin, *Rynchophrya* Collin, *Dactylophorya* Collin, and *Ophryodendron* Claparede and Lachmann occur on the integument of insects in a phoretic association. For detailed descriptions of these forms, see Kudo (1966). When numerous, they may impair locomotion or respiration and thus be pathogenic.

3. Cilia present in both juvenile and mature stages; tentacles absent—**4**

4. Mature stages with oral cilia only; somatic cilia lacking; often stalked (Subclass Peritrichia)—**5**

4. Mature forms with cilia over their entire bodies—Subclass Holotrichida. The only known insect pathogens in this order belong to the genus *Tetrahymena* Furgason (Figures 64, 65). At least three species occur in the body cavity of insects, especially dipterous larvae. See Corliss (1960) for a discussion of this genus.

5. Mobile forms without stalk—Suborder Mobilina. Representatives of the genus *Urceolaria* Lamarck (*Trichodina* Ehrenber) and others may occur on the integument of aquatic insects.

5. Predominantly sessile, with contractile or noncontractile stalk—Suborder Sessilina. Representatives of the genera *Vorticella* L. (Figure 66) (see Noland and Finley, 1931), *Epistylis* Ehrenberg (see Nenninger, 1948), and others occur on the integument of aquatic insects. Some physical impairment may occur when large numbers are present.

6. Motile stages showing amoeboid movement (occur within host cells during the early stages of infection); resting stages are cysts in the malpighian tubules and/or gut of the host (Figure 69); flagella absent in all stages (Superclass Sarcodina, Order Amoebida)—**7**. The two species in couplet 7 are the most common entomogenous pathogens of this order. Two other species that are less frequently encountered are *Malpighiella refringens* Minchin from fleas and *Dobellina mesnili* [Keilin]

from gnats. The insect-parasitic amoebae are reviewed by Lipa (1963) and W. M. Brooks (1974).

6. Motile stages with one or more flagella; small ovoid (leishmanial) nonmotile stages may also be present in the host's gut— Superclass Mastigophora (Figures 67, 68). In general, flagellates are only occasionally encountered in insects and are rarely responsible for debilitation or mortality. Members of the orders Retortamonadida Grassé, Trichomonadida Kirby, and Hypermastigida Grassé and Foa develop in the gut of insects, especially roaches and termites. Most of the parasitic genera of the order Kinetoplastida (most common in Diptera and Hemiptera) occur in the suborder Trypanosomatina and include some medically important species. The genera *Crithidia* Leger (Figure 68), *Blastocrithidia* Laird, *Herpetomonas* Kent, *Rhynichoidomonas* Patton, and *Leptomonas* Kent habitually live in the alimentary tract of invertebrates (see Wallace, 1966). Occasionally they may break through the intestinal wall and invade the hemocoel and other internal tissues of the host. Smirnoff (1974) reported reduced viability in sawfly larvae infected with *Herpetomonas* sp. Members of *Phytomonas* Donovan are plant pathogens which are vectored by insects. The two genera *Leishmania* Ross (common in *Phlebotomus* flies in Europe, Asia, and Africa) and *Trypanosoma* Gruby (common in blood-sucking insects in Africa, South America, and tropical islands) occur in the host gut and are carried by blood-sucking insects to vertebrates, often causing serious diseases. See Kudo (1966) and Wallace (1966) for accounts of the entomogenous trypanosomatids.

7. Infects the malpighian tubules of adult honeybees— *Malpighamoeba mellificae* Prell.

7. Infects the malpighian tubules and gut of locusts and grasshoppers—*Malameba locustae* (King and Taylor) (Figure 69). See Lipa (1963) and W. M. Brooks (1974) for a discussion of this species.

8. Spores variable in shape, usually not round or navicular; containing a polar filament (Subphylum Cnidospora)—9. The polar filament is a threadlike structure coiled within the spore

(Figures 81, 82). The filament may be used to initiate penetration into the host cells. Several methods can be used to extrude the polar filament mechanically on a microscope slide (Ishihara, 1967). One can apply pressure to the coverslip with the thumb or blunt object or use chemicals such as acetic acid, ammonia, glycerine, hydrogen peroxide, or iodine. Rarely, the filament can be demonstrated within the spore using Giemsa stain.

8. Spores mostly round or navicular; not containing a polar filament (Subphylum Sporozoa)—**29**

9. Spores containing an elongate filament and three spherical sporoplasms; filament not attached to the spore after extrusion—*Helicosporidium*. *H. parasiticum* Keilin (Figure 70) was considered an ascomycete at one time, and there is still some question regarding its taxonomic position. See Kellen and Lindegren (1974) for a discussion of this pathogen, which occurs in fat body and nerve tissue of insect larvae.

9. Spores with only a polar filament in the form of a thin, coiled tube (Figures 81, 82); filament remains attached to the spore (Class Microsporidea)—**10.** These are highly specialized parasites which occur widely in insects and other animals. Identification is based mainly on spore characteristics. For a review of these groups, see Thomson (1960), Weiser (1961), Tuzet *et al.* (1971), and W. M. Brooks (1974).

10. Spore with a short polar filament, lacking a polaroplast, many spores produced within a thick-walled sporocyst or a thin membrane or both (rare)—**11**

10. Spore with a normal polar filament; with polaroplast (Figure 82); sporocyst present (as a pansporoblastic membrane) or absent (Order Microsporida)—**12.** This order is currently undergoing revision by several authorities, and many new taxa are being proposed. Many of the diagnostic characters separating newly proposed genera are based on the ultrastructure of the spores. Unless otherwise noted, all genera in this order are placed in the family Nosematidae.

11. Spores develop within a thick-walled envelope; parasite without any special relationship to the host nucleus; found in *Sciara* fly larvae—*Hessea* Ormières and Sprague (Family Hesseidae). See Ormières and Sprague (1973) for a discussion of this genus.

11. Spores may develop within a thick-walled cyst or a thin envelope; parasite develops in intimate contact with the host cell nucleus; parasites of Coleoptera—*Chytridiopsis* Schneider (Family Chytridiopsidae). See Sprague *et al.* (1972) for a review of this genus.

12. Spores rod- or sausage-shaped, more than twice as long than wide—**13**

12. Spores round to elliptical, not more than twice as long as wide—**19**

13. Spores tubular, generally 7–10 times longer than wide; spores with a needle-like structure (manubrium) running through their axis (Family Mrazekidae)—**14**

13. Spores curved or bent; if tubular, then at most 5–6 times longer than wide; manubrium absent (Family Cougourdellidae)—**15**

14. Spores with a caudal mucron (appendage or process)— *Mrazekia* Léger and Hesse (Figure 71). See Weiser (1961) for a discussion of this genus.

14. Spores without a caudal mucron, appearing similar to large *Bacillus* rods—*Bacillidium* Janda (Figure 72). See Weiser (1961) for a discussion of this genus, which is considered by some as a synonym of *Mrazekia* Léger and Hesse.

15. Spores curved or bent—**16**

15. Spores rod shaped—**17**

16. Spores C-shaped, with a single bend—*Toxoglugea* Léger and Hesse (Figure 73). See Poisson (1941) for a discussion of this genus.

16. Spores S-shaped, with a double bend—*Spiroglugea* Léger and Hesse (Figure 74). See Weiser (1961) for a discussion of this genus, which is considered by some as a synonym of *Toxoglugea* Léger and Hesse.

17. Spores of uniform thickness throughout—*Octosporea* Flu (Figure 75). See Weiser (1961) for a discussion of this genus.

17. Spores swollen at one end—**18**

18. Swollen end of spore uniformly curved, opposite end blunt—

Cougourdella Hesse (Figure 76). See Hesse (1935) for an account of this genus.

18. Swollen end of spore curved, opposite end tapering, elongate pyriform in shape—*Pyrotheca* Hesse (Figure 77). See Hesse (1935) for an account of this genus, which may simply represent a morphological modification of *Cougourdella* Hesse.

19. Spores containing a terminal mucron or appendage longer than the spore itself; lateral gelatinous processes or a crest may be present—*Caudospora* Weiser (Figure 78) (Caudosporidae). See Weiser (1961) for a discussion of this genus.

19. Spores without a mucron or gelatinous processes—**20**

20. Spores bearing a crest or crown shaped structure at one end— *Weiseria* Doby and Saguez (Figure 79). See Doby and Saguez (1964) and Jamnback (1970) for discussions on this genus.

20. Spores lacking a crest—**21**

21. Spores formed singly or if in pairs, then they soon separate to appear singly; not in groups of pansporoblasts—*Nosema* Nägeli (Figures 80, 81). See Kramer (1964) for a discussion of a *Nosema* infection. *Perezia* Léger and Dubosq is considered by some as a synonym of *Nosema*.

21. Spores formed in pansporoblasts—**22**

22. Pansporoblasts not containing a constant number of spores—**23**

22. Pansporoblasts containing a constant number of spores—**24**

23. Pansporoblasts containing 4 or 8, or 8 and 16 spores— *Stempellia* Léger and Hesse. See Hazard and Savage (1970) for a discussion of a *Stempellia* infection.

23. Pansporoblasts containing a varied number of spores (generally over 16)—*Pleistophora* Gurley (Figures 82, 83). For descriptions of *Pleistophora* spp., see Canning (1957) and Clark and Fukuda (1971).

24. Pansporoblasts with two spores—**25**

24. Pansporoblasts with more than two spores—**26**

25. Spores side by side, separating under pressure—*Glugea* Thelohan (Figure 84). *Perezia* Léger and Dubosq is considered by some as a synonym of *Glugea*. For descriptions of these genera, see Kramer (1959) and McLaughlin (1969).

25. Spores end to end, never separating—*Telomyxa* Léger and

Hesse (Figure 85). See Poisson (1941), Weiser (1961), and Codreanu (1963) for discussions of this genus.

26. Pansporoblasts with four spores—*Gurleya* Doflein. See Weiser (1961) for an account of this genus.

26. Pansporoblasts with eight or more spores—**27**

27. Pansporoblasts with eight spores—*Thelohania* Henneguy (Figure 86). For a discussion of this genus, see Hazard and Weiser (1968). Hazard and Anthony (1974) described *Parathelohania* Codreanu as differing from *Thelohania* on the basis of its possessing 2 distinct life cycles. The normal cycle produces 8 oval spores in a membrane in host larvae (Culicidae) and an alternative cycle produces a variable number of elongate and slightly curved spores without a membrane in adult hosts.

27. Pansporoblasts with 16 (rarely 32) spores—**28**

28. Pansporoblasts with four needle-like structures in the wall—*Trichoduboscqia* Léger (Figure 87). See Léger (1926) for an account of this species.

28. Pansporoblasts without needle-like structures—*Duboscqia* Perez (Figure 88). See Kudo (1942) for a discussion of this genus.

29. Spores small, nearly spherical; borne in thin-walled pansporoblasts similar to microsporidans—Order Haplosporida (Figure 89). Representatives of the genera *Coelosporidium* Mesnil and Marchoux (see Sprague, 1940), *Haplosporidium* Caullery and Mesnil (see Sprague, 1963), *Coelomycidium* Mesnil et Marchoux (see Caullery and Mesnil, 1905), *Myiobium* Swellengrebel (1919), *Myrmecisporidium* Hölldobler (1930), *Mycetosporidium* Léger and Hesse (1905), and *Nephridiophaga* Ivanic (see Woolever, 1966) are found in various tissues in a range of insects. However, they are rarely encountered on routine diagnoses and usually do not cause disease.

29. Spores variable in size and shape; generally larger than those of the Microsporida and Haplosporida; borne in thin or thick walled cysts—**30**

30. Cysts protruding from the stomach wall of blood sucking insects as small tumors (40–60 μm in diameter) (Figure 90); cyst ruptures to produce elongate sporozoites that migrate to the

salivary glands of the host and are infective to vertebrates; motile worm-shaped zygotes may occur in the body cavity—Suborder Haemosporina. Members of this group are heteroxenous parasites and require both invertebrate and vertebrate hosts to complete their development. The malarial parasites (*Plasmodium* spp.) (Figure 90) occur in this suborder, along with *Haemoproteus* Kruse in pigeons and *Leucocytozoon* Danilewsky from ducks. For representatives of these genera, which all undergo partial development in blood-sucking insects, see Kudo (1966). Certain pathologies are caused in the insect host during development of these parasites, but they are rarely lethal and have been little studied.

30. Cysts not protruding from the stomach wall; monoxenous forms parasitizing invertebrates only—**31**

31. Mature trophozoites (vegetative or developing forms) large, extracellular, club-shaped, elliptical (in gut), or spherical (in body cavity); usually septate (divided into a protomerite and deutomerite); spores generally small and inconspicuous; found in the gut or body cavity (Subclass Gregarinia)—**32**

31. Mature trophozoites small; mostly intracellular; nonseptate; spores or oocysts large—**36**

32. Mature trophozoites septate; attached to gut wall or free in gut lumen—Order Eugregarinida, Suborder Cephalina (Figure 91). There are many genera parasitic in insects (especially Diptera, Orthoptera, and Coleoptera), which are treated by Kudo (1966). Since most species are considered harmless to their hosts, they will not be treated further here. For a description of a representative species in insects, see Canning (1956).

32. Mature trophozoites aseptate; attached to gut wall or in body cavity (Order Eugregarinida, Suborder Acephalina) (Figure 92)—**33**

33. Cysts spherical—**34**

33. Cysts sausage-shaped—*Allantocystis* Keilin. Keilin (1920) described one species from the fly, *Dasyhelea obscura*.

34. Cysts found in gut lumen or malpighian tubules; gamonts elongate, motile; commonly found in mosquitoes—*Lankesteria*

Mingazzini (Figure 93). See Sanders and Poinar (1973) for a discussion of a species in this genus.

34. Cysts found in body cavity (rarely gut lumen); gamonts spherical—**35**

35. Spores round or oval; found in Orthoptera (especially cockroaches and crickets)—*Diplocystis* Kunstler. See Kudo (1966) for an account of this genus.

35. Spores biconical (pointed at both ends); found in Coleoptera—*Monocystis* Stein. See Kudo (1924) for an account of this genus. The closely related genus *Enterocystis* Zwetkow (see Codreanu, 1940) occurs in ephemerids.

36. Spores oval or navicular, containing 8 sporozoites; found in gut epithelium, malpighian tubules or fat body (Order Neogregarinida)—**37.** The two most commonly occurring genera are listed below. Others are discussed by Weiser and Briggs (1971).

36. Spores mostly spherical; within resting body; containing a variable number of sporozoites, but rarely 8; found in hemolymph, fat body, and occasionally gut epithelium (Subclass Coccidia, Order Eucoccida)—**38**

37. Cysts containing 8 spores; common in *Tribolium* sp.—*Farinocystis* Weiser. See Dissanaike (1955) for an account of this genus.

37. Cysts containing 1–2 spores; common in stored-product insects—*Mattesia* Naville (Figure 94). For discussion of a *Mattesia,* see McLaughlin (1965).

38. One sporozoite in each spore; spores bivalve; gametocysts similar—*Barrouxia* Schneider. See Kudo (1966) for a discussion of this genus. *B. ornata* Schneider occurs in the gut of *Nepa cinerea* L.

38. Two to six sporozoites in each spore; spores not bivalve; gametocytes dissimilar—**39**

39. Spore walls not formed; sporozoites occur within the oocyst wall—*Legerella* Mesnil. See Vincent (1927) for a discussion of this genus which infects the malpighian tubules of fleas and the beetle *Hydroporus palustris* L. (Dytiscidae).

39. Spore walls present—**40**

40. Cysts with three spores, each spore with 4–6 or more sporozoites—*Chagasella* Machado. See Machado (1913) for a discussion of this genus, which is found in the intestinal epithelium of the bug, *Dysdercus ruficollis* L.

40. Cysts with more than 3 spores—**41.**

41. Cysts with four spores—*Ithania* Ludwig. *Ithania wenrichi* Ludwig occurs in the intenstinal epithelium of tipulid larvae (Ludwig, 1947).

41. Cysts with usually more than 4 spores; each with 2 sporozoites—*Adelina* Hesse (Figure 95). See Yarwood (1937) for an account of a species in this genus. They are commonly encountered in fat body and other tissues in a variety of insects.

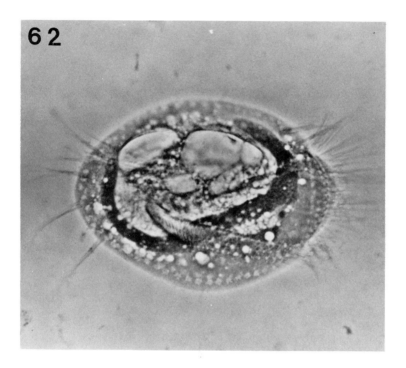

FIGURE 62

A representative of the class Ciliatea (*Euplotes* sp.). ×600

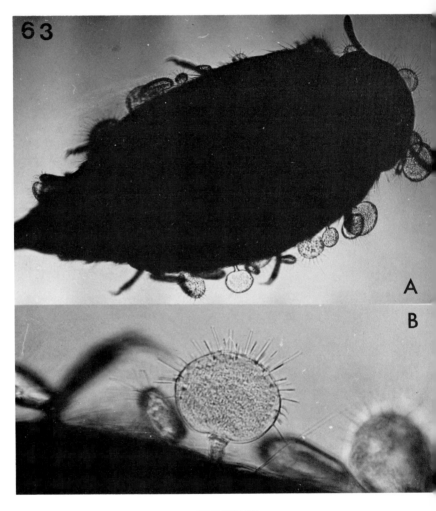

FIGURE 63

A representative of the class Suctorea from a water beetle. (A) ×34, (B) ×340.

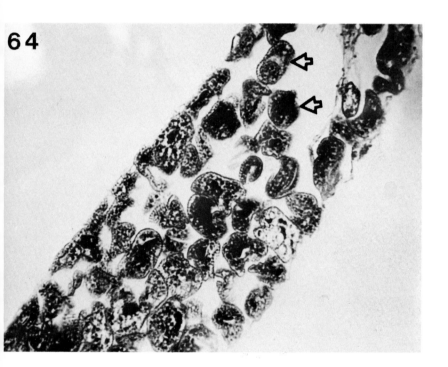

FIGURE 64

Fixed preparation of *Tetrahymena* sp. (arrows) filling the hemocoel of *Aedes sierrensis* (courtesy of D. Sanders). ×200

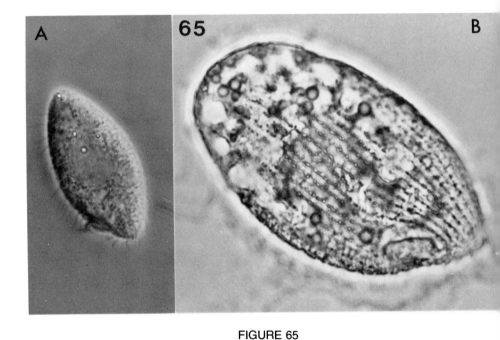

FIGURE 65

Tetrahymena pyriformis. (A) Phase contrast of living individual (×1000). (B) Silver nitrate stain of fixed individual (×2000), reproduced @ 90%.

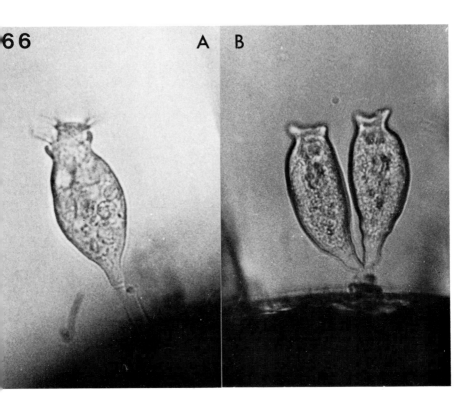

FIGURE 66

ɔrticella sp. from a chironomid larva. (A) Expanded (×970). (B) Contracted (×970), reproduced @ 90%.

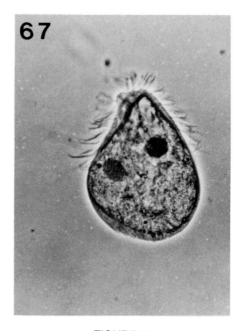

FIGURE 67

A representative of a flagellate from the gut
of a termite. ×260

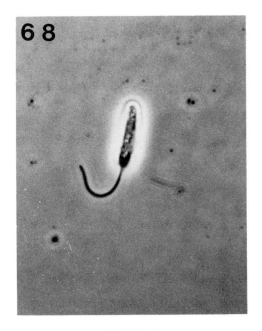

FIGURE 68

The flagellate, *Crithidia fasciculata* (courtesy
of W. Balamuth). ×1800

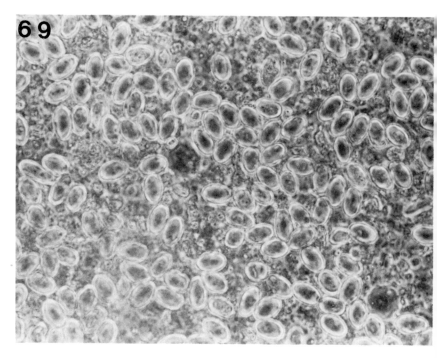

FIGURE 69

Spores of *Malameba locustae* in the malpighian tubules of a grasshopper. ×640

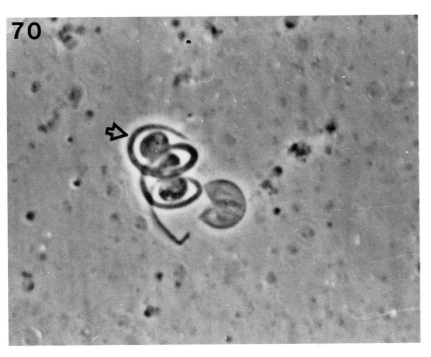

FIGURE 70

Helicosporidium parasiticum. Note elongate detached polar body (arrow)
(courtesy of W. Kellen). ×1700

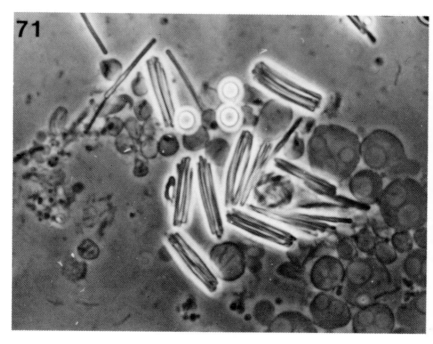

FIGURE 71

Spores of *Mrazekia* sp. from *Cladotanytarsus* sp. (courtesy of E. Hazard). ×1320

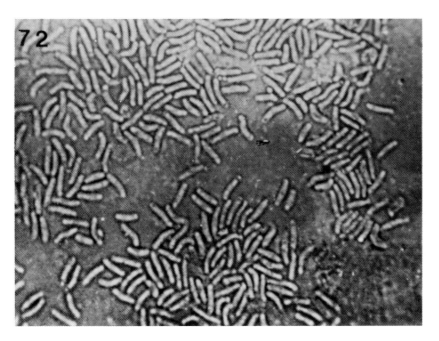

FIGURE 72

Spores of *Bacillidium* sp. from *Chironomus plumosus* (courtesy of J. Weiser).
×600

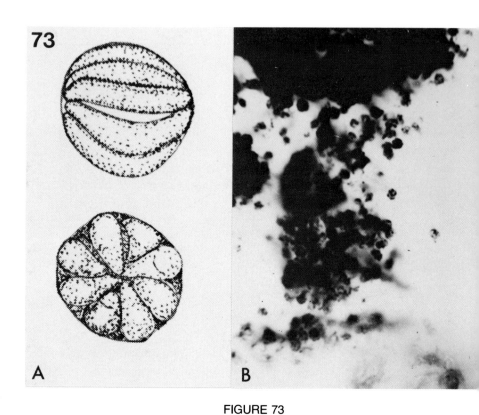

FIGURE 73

Spores of *Toxoglugea* sp. (A) Pansporoblast (after Weiser, 1961). (B) Pansporoblast an spores (courtesy of J. Weiser). ×900, reproduced @ 90%.

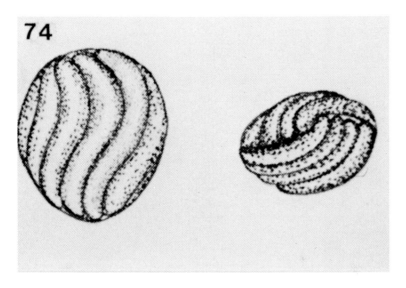

FIGURE 74

Sporoblast of *Spiroglugea octospora* (after Weiser, 1961).

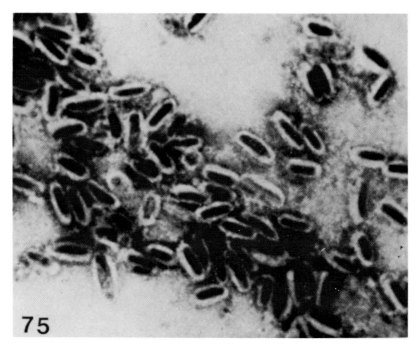

FIGURE 75

Spores of *Octosporea viridanae* (courtesy of J. Weiser). ×1000

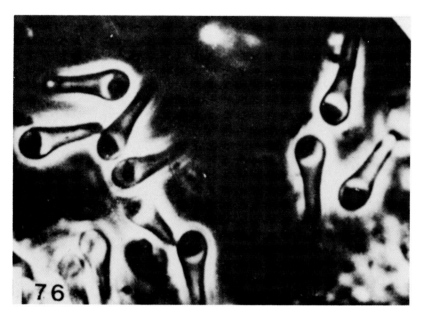

FIGURE 76

Spores of *Cougourdella rhyacophilae* (after Baudoin, 1969). ×6000

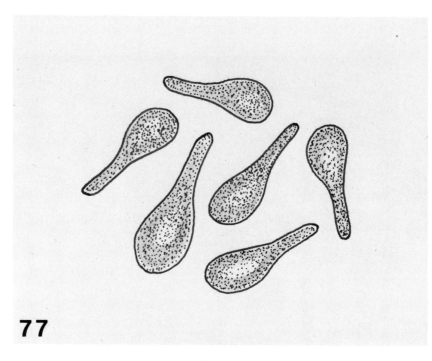

77

FIGURE 77

Spores of *Pyrotheca* sp.

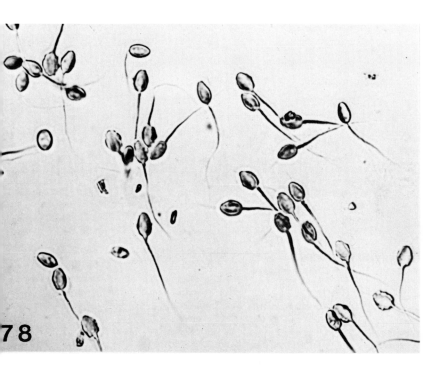

FIGURE 78

Spores of *Caudospora simulii* (courtesy of J. Weiser). ×1000

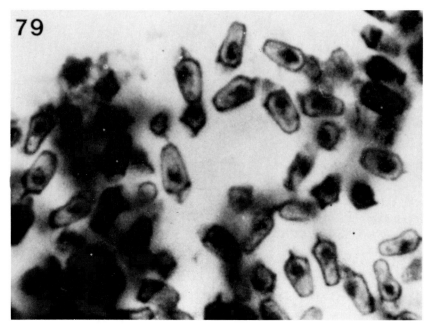

FIGURE 79

Spores of *Weiseria laurenti* (courtesy of J. Weiser). ×1000

FIGURE 80

Spores of *Nosema* sp. (A) Wet mount (×600). (B) Giemsa stain (×1600).

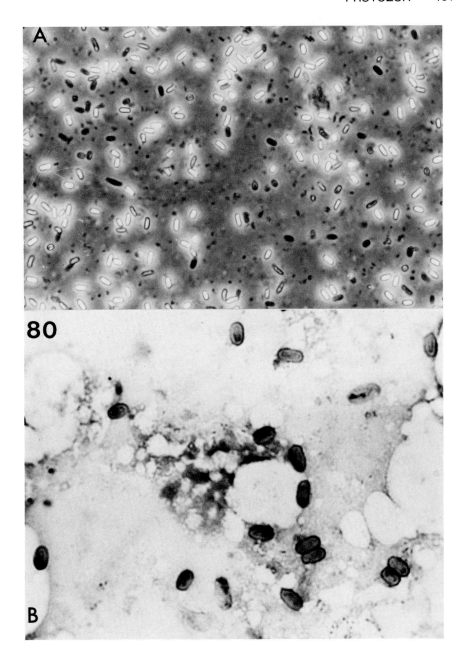

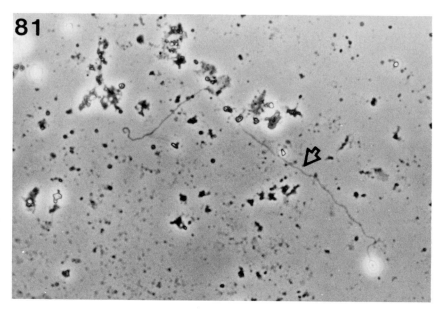

FIGURE 81

Extruded polar filament (arrow) from a spore of *Nosema* sp. ×640

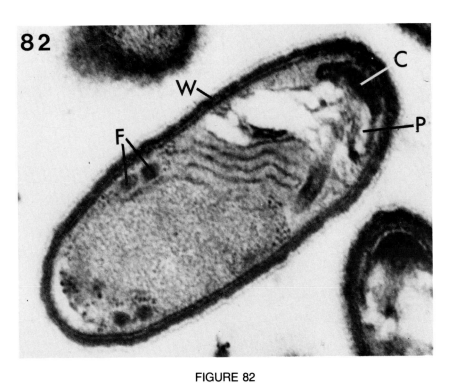

FIGURE 82

Electron micrograph of a *Pleistophora* sp. spore. C, polar cap; F, section of polar filament; W, spore wall; P, polaroplast (courtesy of D. Sanders). ×84,900

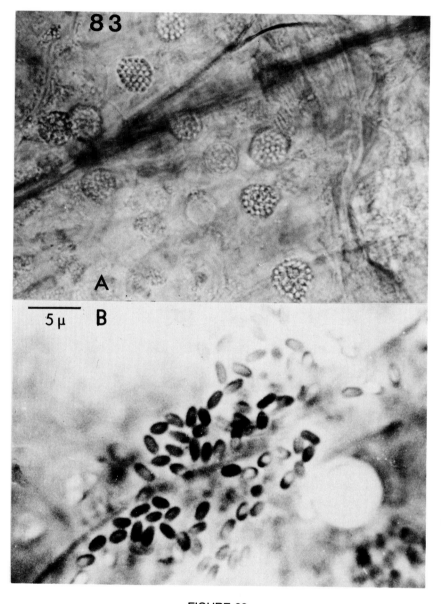

FIGURE 83

(A) Pansporoblasts and (B) spores of *Pleistophora* sp. (courtesy of D. Sanders).

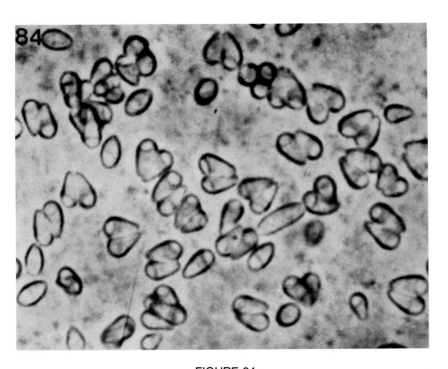

FIGURE 84

Spores and pansporoblasts of *Glugea trichopterae* (courtesy of J. Weiser).
×1200

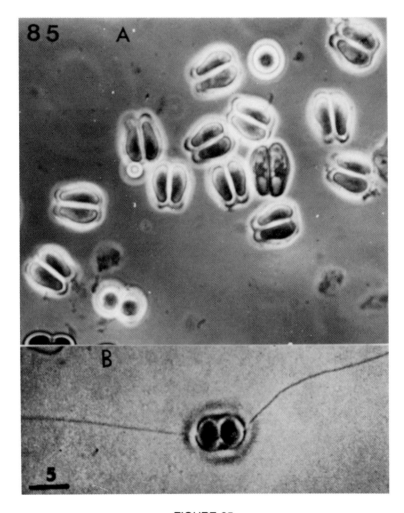

FIGURE 85

Telomyxa sp. (A) Diplospores from a helodid beetle larva (courtesy of E.
Hazard) (×1370). (B) Diplospore with extruded filaments
(after Codreanu and Vavra, 1970) (×2000).

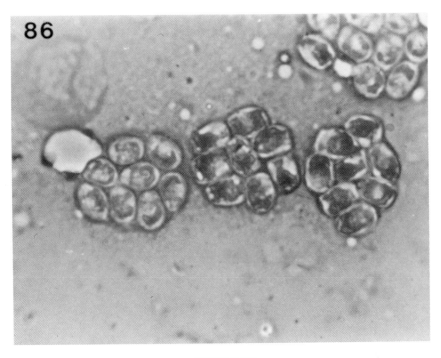

FIGURE 86

Pansporoblasts of *Thelohania californica* from *Culex tarsalis*
(courtesy of D. Sanders). ×1600

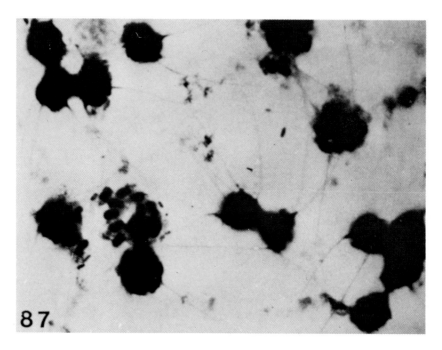

FIGURE 87

Pansporoblasts (Giemsa stained) of *Trichoduboscqia epeori*
(courtesy of J. Weiser). ×500

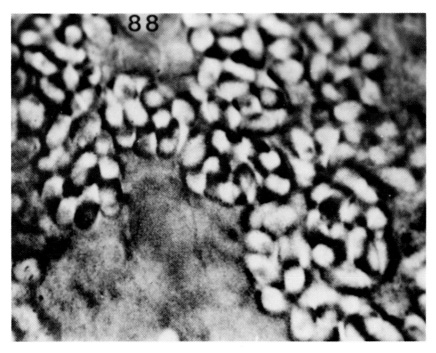

FIGURE 88

Pansporoblasts of *Duboscqia legeri* (after Kudo, 1942). ×2400

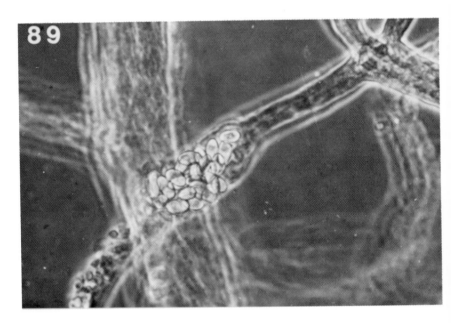

FIGURE 89

A representative of the Haplosporida in the ganglion of *Tenebrio molitor* (courtesy of W. Kellen). ×900

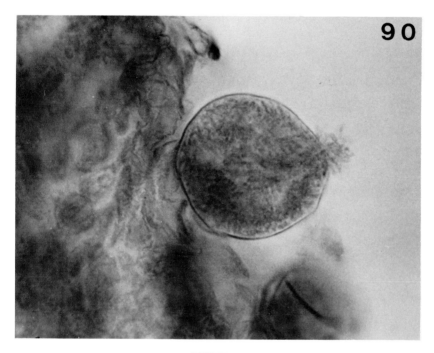

FIGURE 90

Rupturing oocyst of *Plasmodium circumflexum* on the gut wall of *Culiseta morsitans* (courtesy of M. Laird and E. Greiner). ×900

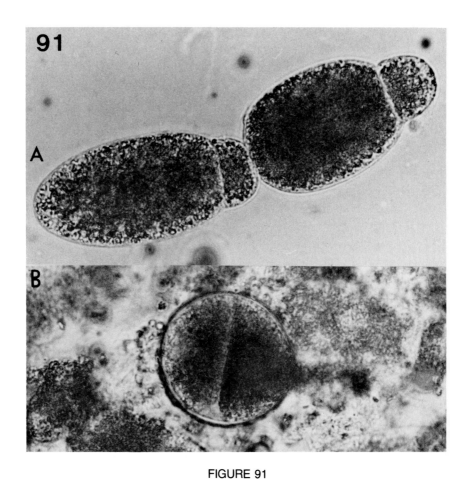

FIGURE 91

(A) Mature trophozoites and (B) cyst of a member of the Cephalina removed from the gut of the beetle, *Carpophilus* sp. ×640

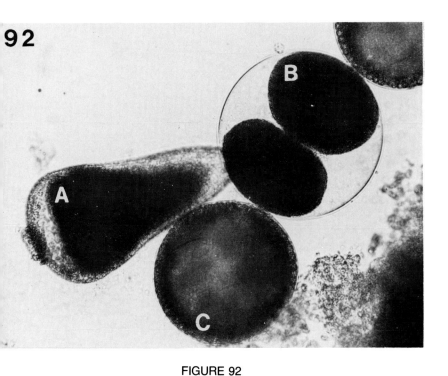

FIGURE 92

A) Mature trophozoite, (B) encysted gamonts, and (C) gamontocyst of a member of the Acephalina removed from a flea (courtesy of B. Nelson). ×640

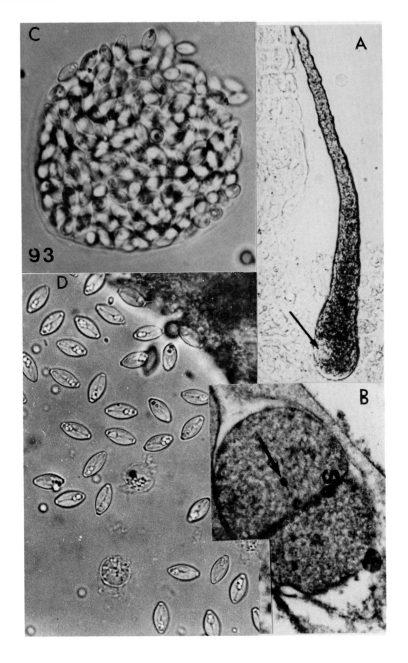

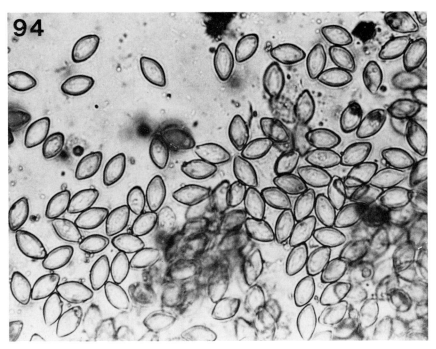

FIGURE 94

Spores of *Mattesia* from the wasp, *Bathyplectes* sp. ×680

FIGURE 93

Lankesteria clarki in *Aedes sierrensis:* (A) elongate gamont in midgut (×800);
(B) gamontocyst in malpighian tubule (×500); (C) mature gamontocyst with spores
(×650); (D) mature spores (×1000) (courtesy of D. Sanders), reproduced @ 85%.

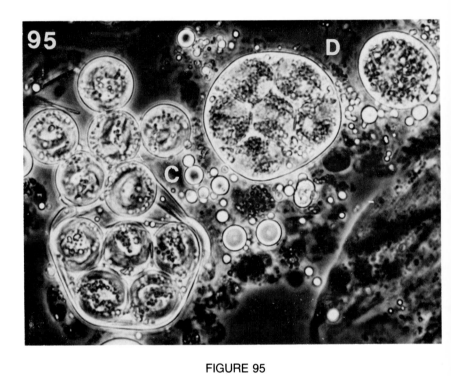

FIGURE 95

Adelina sp. from *Trogoderma* showing developing cysts (D) and a mature cyst releasing spores (C). ×1000

RICKETTSIAS

INTRODUCTION

The rickettsias associated with insects may be intracellular, epicellular (living on the surface of cells), or extracellular; and they may have phoretic, mutualistic, or pathogenic relationships with their hosts. For example, *Rochalimaea quintana* (Schmincke), the causal agent of trench fever in man, grows extracellularly in the gut lumen of the human louse, *Pediculus humanus*. On the other hand, *Rickettsia prowazekii* da Rocha-Lima, the causal agent of typhus fever in man, is pathogenic to its vector, *P. humanus*. Rickettsias which are considered essential for development and reproduction of the host are *Symbiotes lectularius* in mycetomes of the common bedbug (*Cimex lectularius*) and *Blattabacterium* sp. in mycetocytes of cockroaches. Rickettsias belonging to the genus *Rickettsiella* are pathogenic to their insect hosts.

Hosts of rickettsias range from trematodes to man. Many of the vertebrate forms transmitted by arthropods are pathogenic to both vector and vertebrate. Because of their wide host range, some consider all rickettsias as potentially dangerous to humans and they should be handled with care. The number of insect hosts and/or vectors of rickettsias is small at present. However, it may be shown that the majority of insects carry these microorganisms somewhere in their bodies.

TAXONOMIC STATUS

The infective stages of rickettsias are minute structures, about the size of granulosis virus inclusion capsules (0.2–0.6 μm). The majority are rod-shaped, coccoid, or pleomorphic and are bound by a typical cell wall containing muramic acid. They do not possess flagella and are Gram negative (with one exception). They contain both RNA and DNA and possess metabolic enzyme systems which can be inhibited by chemotherapeutic and antibiotic agents. With the exception of those few forms that multiply epicellularly or extracellularly, the rickettsias multiply within their host cells.

The eighth edition of *Bergey's Manual of Determinative Bacteriology* divides the class Rickettsias into two orders, the Rickettsiales and the Chlamydiales. Only members of the order Rickettsiales are known to be associated with insects; therefore, the Chlamydiales (see Page, 1974) will not be considered further here. The taxonomic scheme for the insect-associated rickettsias in the order Rickettsiales, as presented in *Bergey's Manual,* is outlined in Table 1.

LIFE CYCLE

Infective stages of insect pathogenic rickettsias invade the host cell and develop into dividing vegetative stages. This division may be by simple binary fission or through the production of secondary cells which then produce smaller cells that develop into the thick-walled infective stages. Several different morphological types of cells often are encountered during rickettsial development. There are small, tubular infective stages, larger, often pleomorphic, rod-shaped vegetative stages and spherical cells. The latter two often occur in vesicles or vacuoles which frequently harbor crystal-like inclusion bodies. The most commonly encountered pathogens of insects belong to the genus *Rickettsiella* and develop intracellularly in various tissues, especially the fat body.

Infection of fleas or lice with *Rickettsia prowazekii* and *R. typhi* is initiated when the insects feed upon an infected vertebrate. In the insect, the *Rickettsia* multiplies in the cytoplasm of the gut

Table I. Systematic Arrangement for the Order Rickettsiales Associated with Insects

Family	Tribe	Genus	Species	Insect association
Rickettsiaceae	Rickettsiae	*Rickettsia*	*prowazekii* da Roche-Lima	Vector and pathogen
			typhi (Wolbach and Todd) Philip	Vector and pathogen
		Rochalimaea	*quintana* (Schmincke) Kreig	Vector
	Wolbachieae	*Wolbachia*	*pipientis* Hertig	Mild pathogen
			melophagi (Nöller) Philip	Commensal
		Symbiotes	*lectularius* (Arkwright *et al.*) Philip	Mutualist
		Blattabacterium	*cuenoti* (Mercier) Hollande	Mutualist
		Rickettsiella	*popillae* (Dutky & Gooden) Philip	Pathogen
Bartonellaceae		*Bartonella*	*bacilliformis* (Strong *et al.*) Strong, Tyzzer, and Sellards	Vector
Anaplasmataceae		*Anaplasma*	*marginale* Theiler	Phoretic vector
			ovis Lestoquard	Phoretic vector
		Haemobartonella	*muris* (Mayer) Tyzzer and Weinman	Vector
		Eperythrozoan	*coccoides* Shilling	Vector
			ovis Neitz	Vector
			parvum Splitter	Vector

epithelial cells. Heavily infected cells are discharged with the feces and are the source of vertebrate infection.

Rochalimaea quintana is normally ingested by its louse vector while feeding on an infected human. The organism grows epicellularly on the gut epithelial cells (Krieg, 1963) and in the gut lumen (Weiss and Moulder, 1974). However, if *R. quintana* is injected intrahemocoelically into the louse, it will first multiply in the hemolymph and later infect the gut epithelial cells (Krieg, 1963). The feces of the louse become heavily contaminated with *Rochalimaea* cells and are the source of human infection when scratched into the skin.

Weiss (1974) comments that species in the genus *Wolbachia* associated with insects are either extracellular or intracellular but do not develop inside mycetomes and are seldom pathogenic for their hosts. *Wolbachia pipientis* may damage cells of the gonads in the mosquito *Culex pipiens* and the gut epithelium of other insects. The method of transmission from one individual to another is unknown but one might suspect transovarial transmission in *C. pipiens* and ingestion of contaminated food in other insects.

Wolbachia melophagi occurs epicellularly on gut epithelial cells of the sheep ked, *Melophagus ovinus*. There is no evidence of injury to the host, and its almost universal presence suggests a mutualistic relationship. Although no information on transmission is available, the fact that the microorganisms have been found in smears from host eggs (Weiss, 1974) suggests the possibility of transovarial transmission.

The obligate intracellular symbiotes in the genera *Symbiotes* and *Blattabacterium* are transmitted via the host's ovaries to the embryo. Members of the genus *Symbiotes* characteristically occur in mycetomes of bedbugs (*Cimex* sp.) while the genus *Blattabacterium* occurs in mycetocytes in abdominal fat body, germ tissue, and oocytes of cockroaches (M. A. Brooks, 1974).

The normal infection route of the pathogen *Rickettsiella popilliae* is peroral or possibly through wounds. Japanese beetle larvae can be infected by feeding, holding them in soil infested with the organisms, or by injection. Injection is the most effective method since less than six organisms constitute an LD_{50} (Weiss, 1974).

Once the organisms have invaded the hemocoel, infection spreads from the fat body to the blood and other organs.

CHARACTERISTICS OF INFECTED INSECTS

Lice infected with *Rickettsia prowazekii* and *R. typhi* show no initial symptoms, but as the infection progresses, the gut wall is irreparably damaged. A few hours before death the lice turn reddish due to the ingested human blood entering the hemocoel. Fleas infected with *R. typhi* become sluggish and die, but do not turn red.

No particular symptoms occur in insects infected with *Wolbachia pipientis,* since the effects of this organism are extremely mild.

The most striking symptom in insect larvae infected with *Rickettsiella popilliae* is a bluish discoloration within the fat-body cells, thus the common name "blue disease." Infected grubs of *Melolontha* spp. show a change in behavior (Niklas, 1957). While reduced temperatures cause healthy larvae to move deeper into the soil, infected grubs come to the surface.

Large, distinct crystalline bodies accompany rickettsial infections in some insects (Figure 96). The significance of these is unknown.

FACTORS AFFECTING NATURAL INFECTIONS

The insect pathogenic rickettsias fall into two categories: those with and those without alternate hosts. Those with alternate hosts include only *Rickettsia prowazekii* and *R. typhi*. The human louse becomes infected with *R. prowazekii* by feeding on blood from an infected human, and the incidence of infection in the louse population is dependent on the intensity and duration of the rickettsemia in the human population. Similarly, spread of the disease in the human population is dependent on the extent of infection in the louse population. Since both hosts are killed by the disease, epizootics tend to be cyclic and density dependent. Rickettsemias of *R. typhi,* on the other hand, tend to be enzootic because latent infections occur in the

Rodentia, which have a rapid rate of multiplication. Consequently, this disease tends to be less cyclic.

Insect pathogenic rickettsias without an alternation of hosts include *Wolbachia pipientis* and *Rickettsiella popilliae*. Little is known about the epizootiology of *W. pipientis*, probably because the pathogen has so little effect on the host. Natural infections of *R. popilliae* in *Melolontha* larvae probably occur through ingestion of soil particles contaminated with diseased larvae; consequently, the higher the population, the greater is the chance for infection.

METHODS OF EXAMINATION

Material for microscope examination may be prepared either as fresh tissue smears or as fixed histological sections. Since rickettsias are minute microorganisms close to the resolution of the light microscope, unstained smears should be examined under dark field or phase contrast. Rickettsias appear as minute refringent flecks under dark field and as grey flecks under phase contrast. They are often in rapid Brownian movement very similar to the inclusion bodies of granulosis viruses.

The rickettsias can be stained with analine dyes, but the most frequently used stains are the Macchiavello method for tissue smear (blue color indicates positive reaction) and the Giemsa method for fixed specimens (see chapter on Techniques). With the latter method, rickettsias are stained red, while granulosis capsules and bacteria are stained blue. Another technique for differentiating rickettsias from granulosis capsules is Giemsa staining after HCl hydrolysis (Krieg, 1963) (see Techniques). Rickettsias react strongly positive while granulosis capsules are negative.

Multiplication of *Rickettsiella popilliae* results in vacuoles or vesicles filled with rickettsias and crystalline bipyramidal bodies. These vesicles are sometimes referred to as NR bodies because they may be stained with neutral red dye. They are also termed globules of spheroidocytes. The bipyramidal crystals closely resemble albuminoid crystals found in late larval and pupal stages of normal insects. In fixed tissue sections stained with buffered Giemsa, rickettsias, crystals, and globules of spheroidocytes are red, while nuclei, albuminoid crystals, and bacteria are blue.

ISOLATION AND CULTIVATION

Rickettsias rarely can be grown on normal culture media like bacteria. They are best grown in insects inoculated per os, paranal, or intrahemocoelically.

Inocula may be prepared by isolating rickettsias as outlined by Krieg (1963). Infected insects are surface sterilized by treatment with 70% ethanol or ethyl ether, decontaminated in aqueous 0.01% merthiolate, and rinsed in sterile water. The infected organ or insect is homogenized in sterile water. After removal of large particles by low-speed centrifugation (1000*g* for 30 min), the supernatant with the rickettsias can be sedimented by high-speed centrifugation (7500*g* for 1 hr). Small types of rickettsias (*Coxiella* spp. and *Rickettsiella* spp.) may be purified by filtration through membranes of cellulose nitrate with a mean porosity of 0.6 μm.

IDENTIFICATION

To identify insect rickettsias, the type of host association, tissue affinities, gross symptoms, morphology, and staining characteristics should be considered. Because of their size, most rickettsias can be measured accurately only with the electron microscope.

TESTING FOR PATHOGENICITY

Since the normal route of infection for the insect pathogenic rickettsias is through the intestine, pathogenicity tests should be conducted by "per os" inoculation. This can be accomplished by force feeding with a microsyringe or by feeding contaminated food. Inocula can be prepared by the method of Krieg (1963) outlined above. When a vertebrate host is involved, inoculum could consist of blood from the vertebrate.

STORAGE

Rickettsias may be stored by refrigerating, freezing, or freeze drying infected hosts. The preparations should be stored dry since physiological saline and distilled water result in inactivation within

2 to 6 hr at room temperature. The hydrogen ion concentration of the medium is very important and the optimum pH lies near the neutral point (pH 6.4–7.2). Most infected insects can be stored at 4°C for several days. Deep-frozen rickettsias may remain viable for several months; *Rickettsia prowazekii* kept at −20°C remained viable for 8 months, and *Rickettsiella popilliae* maintained at −80°C lasted for more than 3 years (Krieg, 1963). Lyophilization for preservation is practical, but the ampules must be stored frozen, at −10 to −20° C. To protect purified rickettsias from inactivation during freezing, protective colloids such as skimmed milk (pH 7.6), albumin, or peptone can be added.

LITERATURE

A comprehensive account of rickettsias associated with insects is given by Krieg (1963). The most recent taxonomic review of the rickettsias is found in part 18 of the eighth edition of *Bergey's Manual of Determinative Bacteriology* (Buchanan and Gibbons, 1974). A short summary of the rickettsias associated with insects is given by Vaughn (1974), and their possible use as microbial control agents is discussed by Krieg (1971).

KEY TO GENERA AND SPECIES

1. Develop intracellularly; may or may not be pathogenic (extracellular forms may occur due to degeneration of heavily infected cells)—**9**
1. Develop epicellularly or extracellularly; usually nonpathogenic —**2**
2. Develop epicellularly on cells of gut epithelium—**3**
2. Found free in the gut lumen, not attached to epithelial cells—**4**
3. Found in the gut of the human louse, *Pediculus humanus* (L.); develops on epithelial cells and free in the gut lumen; cells 0.2–0.5 by 1.0–1.6 μm—*Rochalimaea quintana* (Schmincke). See Weiss (1974) for a discussion of this genus.
3. Found in the gut of the sheep ked, *Malophaga ovinus* (L.); develops in closely packed rows on epithelial cells; cells 0.3 by

0.6 μm—*Wolbachia malophagi* (Noller). See Weiss (1974) for a discussion of this genus.

4. Associated with flies—**5**
4. Associated with lice on pigs or rodents—**7**
5. Associated with sand flies (*Phlebotomus* spp.)—*Bartonella bacilliformis* (Strong *et al*.). Known to be established only in South and perhaps Central America. See Weinman (1974) for a discussion of this genus.
5. Associated with horse flies (Tabanidae)—**6**
6. Organisms round (0.3–0.4 μm in diameter); 1–8 occur in inclusion bodies (0.3–1.0 μm in diameter)—*Anaplasma* Theiler. See Ristic and Krier (1974) for a discussion of this genus.
6. Cells ring or disc shaped, 0.5–1.0 μm in diameter; inclusion bodies absent—*Eperythrozoon ovis* Neitz. See Ristic and Krier (1974) for a discussion of this species.
7. Associated with the pig louse, *Haematopinus suis* (L.)—*Eperythrozoon parvum* Splitter. See Ristic and Krier (1974) for a discussion of this species.
7. Associated with rodent lice—**8**
8. Associated with the mouse louse, *Polyplax serrata* (Burm.); cells coccoid, 0.4–0.5 μm—*Eperythrozoon coccoides* Schilling. See Ristic and Krier (1974) for a discussion of this species.
8. Associated with the rat louse, *Polyplax spinulosa* (Burm.); cells rod-shaped, 0.1 by 0.3–0.7 μm—*Haemobartonella muris* (Mayer). See Ristic and Krier (1974) for a discussion of this genus.
9. Pathogenic in gut epithelium of lice and fleas—*Rickettsia* da Rocha-Lima (Figures 97, 98)—**10**. See Moulder (1974) for a discussion of this genus.
9. Not found in lice or fleas—**11**
10. Found in the human head louse, *Pediculus humanus capitis* De Geer, and less frequently the body louse, *P. humanus humanus;* causes typhus fever in man—*R. prowazekii* da Rocha-Lima. Infected lice become red a few hours before death, when rickettsias can be found free in the gut lumen and hemocoel.
10. Found in the human louse, *P. humanus,* human flea, *Pulex*

irritans (L.), rat louse, *Polyplax spinulosa* (Burm.), and the rat flea, *Xenophylla cheopis* (Roth.). Causes murine (endemic) typhus in man (incidental host) and rats and other rodents (primary reservoir). Cells distinguishable from *R. prowazekii* only by serological techniques—*R. typhi* (Wolbach and Todd) (known also as *R. mooseri*).

11. Nonpathogenic mutualists found in special cells (mycetocytes) or special organs (mycetomes); especially in bedbugs (*Cimex* spp.) or cockroaches—**12**

11. Not found in mycetocytes or mycetomes; slightly to highly pathogenic—**13**

12. Mutualists found in paired mycetomes in most tissues of bedbugs (*Cimex* spp.)—*Symbiotes* Philip. See M. A. Brooks (1974) for a discussion of this genus.

12. Mutualists found in mycetocytes of the fat body, germ tissue, and oocytes of cockroaches—*Blattabacterium cuenoti* (Mercier). See M. A. Brooks (1974) for a discussion of this genus.

13. Occur in gonads of *Culex pipiens* and gut epithelium of *C. fatigans* and other diptera; slightly to mildly pathogenic—*Wolbachia* Hertig (Figure 99). See Weiss (1974) for a discussion of this genus.

13. Occur in fat body and other tissues of a variety of insects; often highly pathogenic; infected larvae may turn white or bluish-white—*Rickettsiella* Philips (Figures 100–104). Weiss (1974) considers all insect-pathogenic species in this genus as belonging to *R. popilliae* (Dutky and Gooden). In the same paper, he lists the hosts of this rickettsia and Kellen *et al.* (1972) discuss the life cycle and developmental stages of a *Rickettsiella*.

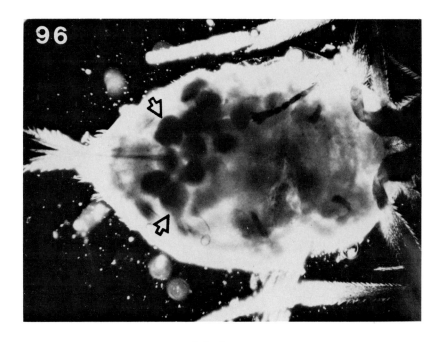

FIGURE 96

Large crystalline bodies (arrows) in the hemocoel of *Bracon hebetor* suffering
from a rickettsial infection (courtesy of W. R. Kellen). ×25

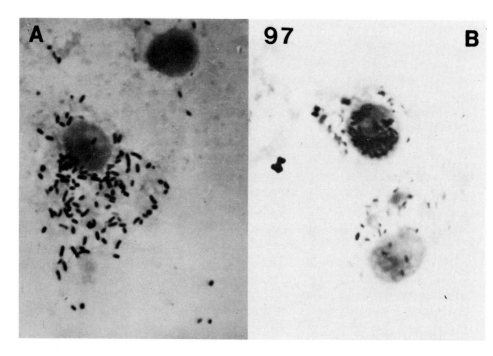

FIGURE 97

Cells of *Rickettsia rickettsii* in (A) the cytoplasm and (B) nucleus of hemocytes of *Dermacentor andersoni* (courtesy of W. Burgdorfer). ×1400

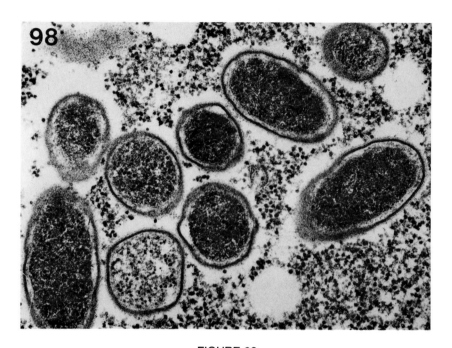

FIGURE 98

Electron micrograph of *Rickettsia rickettsii* in ovarial tick tissue (courtesy of L. P. Brinton and W. Burgdorfer). ×42,000

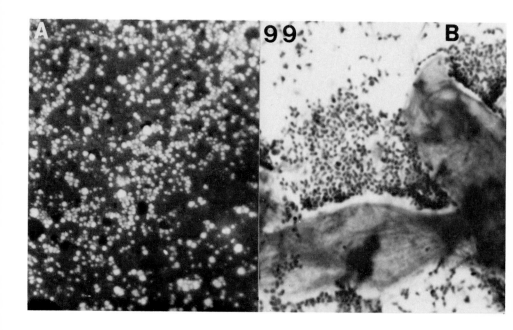

FIGURE 99

Wolbachia cells in ovarial tissue of *Dermacentor andersoni;* (A) coccoid forms (×1300), (B) small rods (×1200) (courtesy of W. Burgdorfer).

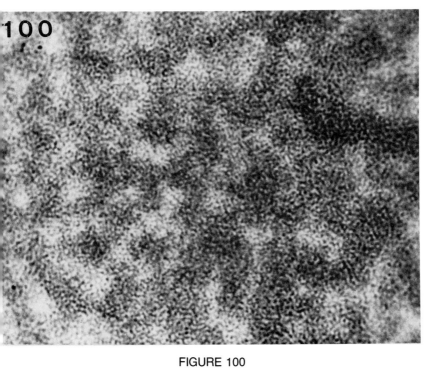

FIGURE 100

Infective stages of a *Rickettsiella* sp. from the larva of a navel orange worm, *Paramyelois transitella.* ×2000

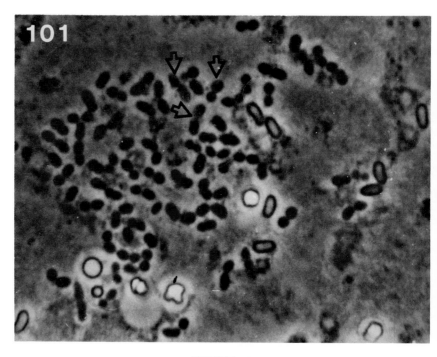

FIGURE 101

Electron micrograph of the developing "daughter cells" (arrows) of a *Rickettsiella* sp. from the navel orange worm (courtesy of W. R. Kellen and D. F. Hoffmann). ×4800

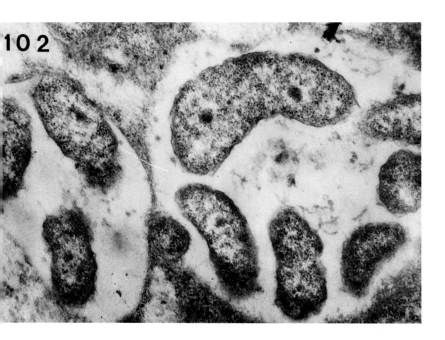

FIGURE 102

Electron micrograph of developing "vesicles" with "secondary cells" in a *Rickettsiella* sp. infection of the navel orange worm (courtesy of W. R. Kellen and D. F. Hoffmann). ×30,250

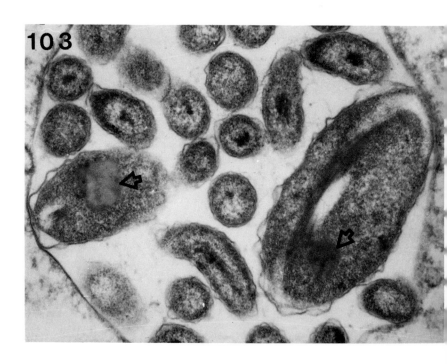

FIGURE 103

Electron micrograph of a large "vesicle" containing "secondary cells" with inclusion bodies (arrows) in a *Rickettsiella* sp. infection of the navel orange worm (courtesy of W. R. Kellen and D. F. Hoffmann). ×38,500

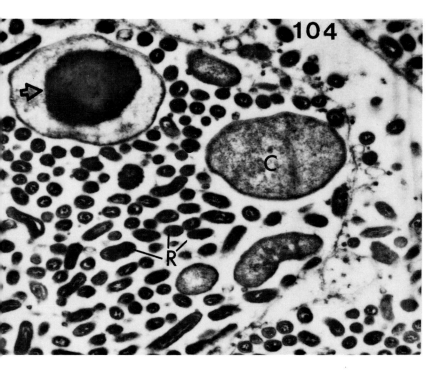

FIGURE 104

Electron micrograph of a *Rickettsiella* sp. infection in *Melolontha melolontha*. Standard type rickettsiae (R) occur in "vacuoles" along with secondary cells (C) and developing crystal (arrow) (courtesy of A. M. Huger). ×25,500

TECHNIQUES

This chapter describes the various techniques and procedures that have been mentioned in the manual. All bacteriological and mycological media discussed can be obtained in dehydrated form from commercial producers such as DIFCO Laboratories, Detroit, Michigan; and BBL, Division of Bioquest, P.O. Box 175, Cockeysville, Maryland. In addition, many are normally stocked by local scientific supply houses.

This portion is divided into four categories; (1) Culture Media, Differential Media, and Ringer's Solution; (2) Bacteriological Tests; (3) Stains and Staining Procedures; and (4) Miscellaneous Techniques. The contents of each section are arranged alphabetically.

CULTURE MEDIA, DIFFERENTIAL MEDIA, AND RINGER'S SOLUTION

AC Medium

AC Medium is a good primary isolation medium possessing unique growth-promoting properties for both aerobic and anaerobic microorganisms. It is recommended as a general culture medium for anaerobes, microaerophiles, and aerobes. The medium contains an agar flux which retards air diffusion and when sealed in snap-cap tubes immediately after autoclaving, it maintains an oxygen gradient ranging from anaerobic at the bottom to aerobic at the top.

Within 24 hr after inoculation, anaerobic organisms will be growing near the bottom, microaerophiles near the center, and aerobic organisms near the top. Facultative organisms will grow throughout the tube. The composition is the following:

```
Beef extract . . . . . . . . . . . . . . . . . . . . . . . . 3 g
Yeast extract . . . . . . . . . . . . . . . . . . . . . . . 3 g
Malt extract . . . . . . . . . . . . . . . . . . . . . . . . 3 g
Proteose peptone No. 3 (Difco) . . . . . . . . . 20 g
Dextrose . . . . . . . . . . . . . . . . . . . . . . . . . . 5 g
Agar . . . . . . . . . . . . . . . . . . . . . . . . . . . . . . 1 g
Ascorbic acid . . . . . . . . . . . . . . . . . . . . . 0.2 g
Distilled water . . . . . . . . . . . . . . . . . . . . 1 liter
```

Mix all ingredients and heat to the boiling point while stirring. Dispense into tubes and autoclave at 15 lb for 15 min. Seal tubes immediately after autoclaving.

Azide Dextrose Broth

Azide dextrose broth is recommended for the isolation and identification of *Streptococcus* spp. Sodium azide is used as an inhibitor for Gram-negative organisms. The medium is reported to be selective for streptococci (Difco Laboratories, 1953), but we found that staphylococci and micrococci, as well as some Gram-positive bacilli, also grow in this medium. *Streptococcus* may be separated from *Staphylococcus* and *Micrococcus* by subculturing on nutrient agar and testing for catalase. *Streptococcus* is catalase negative, while *Staphylococcus* and *Micrococcus* are catalase positive.

```
Beef extract . . . . . . . . . . . . . . . . . . . . . . 4.5 g
Tryptone . . . . . . . . . . . . . . . . . . . . . . . . . 15 g
Dextrose . . . . . . . . . . . . . . . . . . . . . . . . . 7.5 g
Sodium chloride . . . . . . . . . . . . . . . . . . . 7.5 g
Sodium azide . . . . . . . . . . . . . . . . . . . . . 0.2 g
Distilled water . . . . . . . . . . . . . . . . . . . . 1 liter
```

Dissolve the dry ingredients in the water, place 10 ml in each tube, and autoclave at 15 lb for 15 min.

Culture.   The broth is inoculated from a pure culture of the test bacterium, or directly from diseased insect material, and incubated at 30° to 37°C for 24 hr. The resultant growth can be examined microscopically for the presence of cocci. *Streptococcus* spp. usually form long chains in this broth.

Brain–Heart Infusion Agar (BHIA)

Brain–heart infusion agar is a solid medium for the cultivation of fastidious pathogenic bacteria.

Infusion from calf brains . . . . 200 g
Infusion from beef heart . . . . 250 g
Proteose peptone . . . . 10 g
Dextrose . . . . 2 g
Sodium chloride . . . . 5 g
Disodium phosphate . . . . 2.5 g
Agar . . . . 15 g
Water . . . . 1 liter

Mix the dry ingredients with the water and autoclave for 15 min at 15 lb pressure. Pour plates aseptically, when medium is cool enough to handle.

Brain–Heart Infusion Broth (BHIB)

This is a liquid medium recommended for the cultivation of fastidious pathogenic bacteria.

Formula.   The formula for this broth is the same as brain–heart infusion agar, with the agar deleted. Dissolve the dry ingredients in water, dispense in tubes or culture flasks, as needed, and autoclave at 15 lb for 15 min.

Dextrose Agar (DA)

Dextrose agar is recommended as a general all-purpose medium which supports growth of a wide range of bacteria and fungi.

```
Beef extract . . . . . . . . . . . . . . . . . . . . . . . . .  3 g
Tryptose  . . . . . . . . . . . . . . . . . . . . . . . . . .  10 g
Dextrose  . . . . . . . . . . . . . . . . . . . . . . . . . .  10 g
Sodium chloride  . . . . . . . . . . . . . . . . . . . . .  5 g
Agar . . . . . . . . . . . . . . . . . . . . . . . . . . . . . .  15 g
Distilled water  . . . . . . . . . . . . . . . . . . . . .  1 liter
```

Dissolve the dry ingredients in the water, autoclave for 15 min at 15 lb, cool to about 50°C and dispense aseptically into petri dishes.

Dextrose Broth (DB)

Similar to DA, without the agar, and with 5 g of dextrose rather than 10. Dissolve the dry ingredients in water, dispense into tubes or culture flasks, and autoclave at 15 lb for 15 min.

Nutrient Agar (NA)

Nutrient agar is recommended as a general solid medium for the cultivation of many bacteria.

```
Beef extract . . . . . . . . . . . . . . . . . . . . . . . . .  3 g
Peptone  . . . . . . . . . . . . . . . . . . . . . . . . . . .  5 g
Agar . . . . . . . . . . . . . . . . . . . . . . . . . . . . . .  15 g
Water  . . . . . . . . . . . . . . . . . . . . . . . . . . . .  1 liter
```

Mix the dry ingredients with the water and autoclave for 15 min at 15 lb pressure. Pour the plates aseptically, when the medium is cool enough to handle.

Nutrient Broth (NB)

Nutrient broth is a liquid medium recommended for the cultivation of bacteria with general food requirements.

```
Beef extract . . . . . . . . . . . . . . . . . . . . . . . . .  3 g
Peptone  . . . . . . . . . . . . . . . . . . . . . . . . . . .  5 g
Distilled water  . . . . . . . . . . . . . . . . . . . . .  1 liter
```

Dissolve the dry ingredients in the water, dispense into tubes or culture flasks with appropriate closures to protect from contamination, and autoclave at 15 lb pressure for 15 min.

Nutrient Gelatin

Nutrient gelatin is used for the determination of gelatin liquefaction in bacterial identifications.

Beef extract . . . . 3 g
Peptone . . . . 5 g
Gelatin . . . . 120 g
Water   . . . . 1 liter

Dissolve the dry ingredients in the water while stirring and dispense into tubes. Cover with morton caps, cotton stoppers, etc., and autoclave at 15 lb for 15 min. Cool to solidify and store in airtight containers. To test for gelatin liquefaction, the medium is inoculated by a deep stab, and incubated at 20°C, or room temperature.

Sabouraud Dextrose Agar with Yeast Extract (SDA+Y)

SDA+Y is an enriched medium especially suited for the cultivation of fungi, and many of the more fastidious entomopathogenic fungi grow very well on it.

Neopeptone . . . . 10 g
Dextrose . . . . 40 g
Yeast extract . . . . 2 g
Agar . . . . 15 g
Distilled water . . . . 1 liter

Mix the dry ingredients with water and autoclave at 15 lb for 15 min. When cool, but still liquid (about 50°C), pour plates aseptically.

Sabouraud Dextrose Broth with Yeast Extract (SDB+Y)

Similar to SDA+Y but without agar, and the dextrose is reduced to 20 g. Dissolve the dry ingredients in the water, dispense into tubes or culture flasks, and autoclave at 15 lb for 15 min.

Sabouraud Maltose Agar with Yeast Extract (SMA+Y)

This medium is similar to SDA+Y, with maltose substituted for dextrose. It is particularly useful in promoting spore production, where an abundance of spores is necessary for pathogenicity tests, or when difficulty is encountered in obtaining reproductive structures for identification.

Neopeptone . . . . 10 g
Maltose . . . . 40 g
Yeast extract . . . . 2 g
Agar . . . . 15 g

Mix the dry ingredients with the water, autoclave at 15 lb for 15 min, and dispense aseptically into petri dishes.

Streptococcus Faecalis Differential Medium

Streptococcus faecalis is a potential bacterial pathogen often encountered in insects suffering wounds or environmental stress. *Streptococcus* spp. may first be isolated from suspected material by culturing in azide dextrose broth. A specific identification to *S. faecalis* may then be made on the following medium, devised by Meade (1963):

Medium
1. Dissolve in 1 liter of water:
    10 g peptone
    1 g yeast extract
    2 g sorbitol
    5 g tryosine
    12 g agar
2. Autoclave for 10 min at 10 lb pressure (no further autoclaving is necessary).
3. Adjust pH to 6.2.
4. Add 0.01% TTC (2,3,5-triphenyl tetrazolium chloride) and 1.0% thallous acetate.
5. Pour a shallow layer of this medium in sterile petri dishes and allow to set.

6. Add an additional 4 g of trysoine to the remaining medium and keep warm, but near the setting point. Swirl before pouring to keep the tryosine in suspension.
7. Pour a layer of medium with additional tryosine over the original layer in each petri dish, and allow to set.
8. The medium may be stored in airtight plastic bags, or containers under refrigeration until needed.

Test

1. Inoculate a plate of the above medium with a suspension of the suspected bacterium and incubate at 37°C for 4 hr.
2. Transfer culture to 45°C and incubate for 3 days.

Results.   Uniformly dark maroon colonies encircled by clear zones are regarded as *Streptococcus faecalis*.

Ringer's Solution: 1/4 Strength

One-quarter strength Ringer's solution is often preferred in preparing bacterial suspensions, since it is generally more osmotically compatible with bacterial cells than is sterile distilled water. Quarter strength Ringer's may be sterilized by autoclaving or, preferably, by millipore filtration, since autoclaving may cause a precipitate to form.

$$
\begin{array}{ll}
\text{Sodium chloride} & \text{1.12 g} \\
\text{Potassium chloride} & \text{05 g} \\
\text{Calcium chloride} & \text{06 g} \\
\text{Sodium bicarbonate} & \text{02 g} \\
\text{Distilled water} & \text{500 ml}
\end{array}
$$

Tergitol-7 + TTC Agar (T-7+TTC)

Tergitol-7 agar with the addition of TTC (triphenyltetrazolium chloride) is a good medium for the isolation of Gram-negative coliform bacteria. Gram-positive bacteria are inhibited. It is diagnostic for *Escherichia coli* and ***Xenorhabdus nematophilus***. *E. coli* produces yellow colonies surrounded by yellow zones. *X. nematophilus* forms deep blue colonies against a blue background. *Enterobacter* forms red colonies surrounded by yellow zones, while other Enterobacteriaceae and *Pseudomonas* form red colonies against a blue background.

Peptone No. 3 (Difco) .... 5 g
Yeast extract .... 3 g
Lactose .... 10 g
Agar................................ 15 g
Tergitol-7 .... 0.1 ml
Brom thymol blue .... 0.025 g
Water .... 1 liter

Mix the above ingredients, autoclave at 15 lb for 15 min, and allow to cool to about 50°C. Dissolve 40 mg TTC in a few milliliters of water, and add this after millipore filtration to the cooled medium. Aseptically dispense the solution into petri dishes and allow to solidify.

BACTERIOLOGICAL TESTS

Carbohydrate Fermentation Studies

Purple Broth Base. Recommended for the preparation of carbohydrate broths used in fermentation studies of pure bacterial cultures. The concentration of carbohydrate generally employed for testing the fermentation reactions of bacteria is 0.5 or 1.0%. The use of 1.0 rather than 0.5% helps to insure against reversion of the reaction due to depletion of the carbohydrate by some bacteria.

Beef extract........................ 1 g
Proteose peptone.................... 10 g
Sodium chloride .... 5 g
Brom cresol purple .... 0.015 g
Distilled water .... 1 liter
Carbohydrate....................... 10 g

Dissolve the dry ingredients in the water, tube the medium, and autoclave it at 15 lb for 15 min. Simple sugars, such as glucose, may be added before autoclaving. More complex carbohydrates, which are susceptible to hydrolysis during autoclaving, should be sterilized with a minimum amount of heat or added by millipore filtration after sterilization of the base broth. In the latter case, part of the water is used to dissolve the carbohydrate, and the remaining water is used to make the base broth. The base broth is then autoclaved in bulk, and when cool, the dissolved carbohydrate is added aseptically by millipore filtration. The medium is then aseptically dispensed into previously sterilized tubes. To detect the production of gas, a very small tube (Durham tube) is inverted in the culture tube with the broth. Upon autoclaving, the small tube will fill with broth and sink. If gas is produced during fermentation, some of it will collect in the small tube.

Carbohydrates mentioned in this work which can be used with this broth are glucose, arabinose, glycerol, and inositol. Tubed medium is inoculated from a pure culture of the test bacterium and incubated at 30°C for 16–24 hr. A positive reaction is indicated by the development of a yellow color, while uninoculated medium and negative results are indicated by retention of the original purple color of the medium.

Catalase Test

First, culture the organism on nutrient agar or other suitable media, then flood some of the colonies with 3% hydrogen peroxide. The immediate production of gas bubbles indicates a positive reaction.

Cytochrome Oxidase Test

Reagents
1. 1% aqueous N,N-dimethyl-p-phenylenediamine (stable for only 1–2 months under refrigeration).
2. Ethanolic naphthol = 1% alpha naphthol in 95% ethyl alcohol.

Procedure
1. Culture test organism on a suitable solid medium such as nutrient agar.

2. Mix reagents: 3 drops solution 1 to 2 drops solution 2.
3. Flood some of the colonies with this mixture.
4. Colonies of cytochrome-oxidase-positive organisms turn blue.

Hydrogen Sulfide Production: Triple Sugar Iron Agar (TSI Agar)

This medium is recommended for determining several characteristics of Gram-negative enteric bacteria, including the production of hydrogen sulfide. (For details see Difco Laboratories, 1953, p. 166.)

Beef extract	3 g
Yeast extract	3 g
Peptone	15 g
Proteose peptone	5 g
Lactose	10 g
Saccharose	10 g
Dextrose	1 g
Ferrous sulfate	0.2 g
Sodium chloride	5 g
Sodium thiosulfate	0.3 g
Agar	12 g
Phenol red	0.024 g
Distilled water	1 liter

Dissolve the dry ingredients in the water heated to the boiling point. Tube the medium and autoclave at 15 lb for 15 min. The medium may be stored under refrigeration in airtight containers.

Culture. The tubed medium is inoculated by the stab method from a pure culture of the test organism, and incubated at 30–37°C for 16–24 hr.

Results. The production of hydrogen sulfide is indicated by formation of a distinct black color.

Indole Production

1. Culture the bacterium under investigation in 5 ml 1.0% tryptone for 24 hr.
2. Add 0.2–0.3 ml Kovacs' reagent to the above culture.
3. A dark red color in the surface layer constitutes a positive test for indole production. The original yellow color of the solution constitutes a negative test.

Kovacs' Reagent.   Kovacs' reagent is made by dissolving 5 g of *p*-dimethylaminobenzaldehyde in 75 ml of amyl alcohol and adding 25 ml of concentrated hydrochloric acid.

Methyl Red–Voges Proskauer (MR–VP) Test (for Gram-Negative Enteric Bacteria)

This tests for the production of acetylmethylcarbinol or acetoin.

Culture Medium

Peptone	7 g
Dextrose	5 g
Dipotassium phosphate	5 g
Distilled water	1 liter

Dissolve dry ingredients in water, tube at 5 ml per tube, autoclave at 15 lb for 15 min and store in airtight containers under refrigeration.

Culture.   Inoculate the tubed medium from a pure culture of the test organism and incubate at 30°C for 5–7 days.

Test Reagents
    MR Test. Dissolve 0.1 g methyl red in 300 ml 95% ethyl alcohol and dilute to 500 ml with distilled water.
    VP Test. (1) 40% aqueous sodium hydroxide. (2) Solid creatin.

Test
    MR. Add 5 drops of the methyl red solution to 5 ml of a 5–7 day culture. A positive reaction is indicated by a distinct red color, showing the presence of acid. A negative reaction is indicated by a yellow color.

VP. To 5 ml of a 7-day culture, add 25 mg of solid creatin, then 5 ml 40% NaOH. Stopper and shake for 1 min. A positive reaction is shown by development of a red color a few minutes after agitation.

Nitrate Reduction Test

A test for the ability of bacteria to reduce nitrates (NO_3) to nitrites (NO_2).

Nitrate Culture Broth

```
Beef extract . . . . . . . . . . . . . . . . . . . . . . . . . 3 g
Peptone  . . . . . . . . . . . . . . . . . . . . . . . . . . . 5 g
Potassium nitrate . . . . . . . . . . . . . . . . . . . . . 1 g
Distilled water  . . . . . . . . . . . . . . . . . . . . 1 liter
```

Dissolve the dry ingredients in the water, tube, and autoclave at 15 lb for 15 min. Store in an airtight container under refrigeration.

Culture.   Inoculate tubed broth from a culture of the test organism and incubate at 30° to 37°C for 12 to 24 hr.

Test Reagent Solutions
Sulfanilic Acid. Dissolve 8 g sulfanilic acid in 1000 ml 5 N acetic acid.
Alpha-naphthylamine. Dissolve 5 g alpha-naphthylamine in 1000 ml 5 N acetic acid.

Test.   The medium is tested for the presence of nitrites by adding a few drops of each of the above reagent solutions. A distinct pink, red, or rust color indicates the presence of nitrite reduced from the original nitrate. If an organism grows rapidly and reduces nitrate actively, it is suggested that the test for nitrite be performed at an early incubation period since the reduction may be carried beyond the nitrite stage. The test must always be controlled by comparison with a tube of uninoculated medium.

Oxidase Test

1. Culture the bacterium in question on nutrient agar, or another suitable solid culture medium.
2. Flood some of the colonies with a 1% aqueous solution of N,N,N',N-tetramethyl-p-phenylenediamine dihydrochloride. This solution should be stored under refrigeration and discarded after one month, or when it turns blue. (CAUTION—this chemical is extremely toxic; avoid contact with skin or clothing.)
3. Colonies of oxidase-positive bacteria turn blue.

Phenylalanine Deaminase Test

This test is recommended for the separation of the Proteus and Providence groups from other members of the Enterobacteriaceae.

Phenylalanine Agar

Yeast extract	3 g
Dipotassium phosphate	1 g
Sodium chloride	5 g
dl-Phenylalanine	2 g
Agar	12 g
Distilled water	1 liter

Dissolve the dry ingredients in the water by heating to the boiling point. Distribute in tubes, autoclave at 15 lb for 15 min. Allow autoclaved medium to solidify in a slanted position.

Test Reagent. Dissolve 2.0 g ammonium sulfate and 1 ml 10% sulfuric acid in 5 ml half-saturated ferric ammonium sulfate.

Culture. Inoculate slant from a pure culture of the test bacterium and incubate at 30–37°C for 18–24 hr.

Test. Add 5 drops of the test reagent to the slant culture and rotate the tubes to wet and loosen the growth. A characteristic green color develops in Proteus and Providence cultures.

Voges Proskauer (VP) Test for Bacillus spp., Other Gram-Positive Bacteria, and Nonenteric Gram-Negative Bacteria

(For Gram-negative enteric bacteria see Methyl Red–Voges Proskauer (MR–VP) test.) This is a test for the production of acetyl-methylcarbinol or acetoin.

Culture Medium

> Proteose peptone . . . . 7 g
> Sodium chloride . . . . 5 g
> Glucose . . . . 5 g
> Distilled water . . . . 1 liter

Dissolve dry ingredients in water, tube at 6 ml per tube, autoclave at 15 lb for 15 min. Store in airtight container under refrigeration.

Culture. Inoculate tubed medium from a pure culture of the test organism and incubate at 30°C for 5 days.

Test Reagents
1. 5% alpha naphthol in 100% ethyl alcohol
2. 40% aqueous potassium hydroxide

Test. To 6 ml of a 5-day-old culture, add 2.4 ml alpha-naphthol solution and 0.8 ml 40% KOH. Stopper and shake the tube for one minute. Slope the tube and examine at 15 min and 1 hr.

Results. Positive reaction is indicated by the development of a strong red color.

STAINS AND STAINING PROCEDURES

Acid-Fast Staining (Ziehl–Neelsen Method)

Staining Solutions
1. Ziehl's carbol fuchsin

Solution A

> Basic fuchsin (90% dye content) . . . . 0.3 g
> Ethyl alcohol (95%) . . . . 10 ml

Solution B

Phenol . . . . 5 g
Distilled water . . . . 95 ml

Mix solutions A and B

2. Acid alcohol—95% ethyl alcohol with 3% by volume of concentrated HCl.
3. Counterstain

Methylene blue (80% dye content) . . . . . 0.3 g
Ethyl alcohol (95%) 30ml
Distilled water . . . . 100 ml

Staining Procedure
1. Stain dried smear 3–5 min with Ziehl's carbol fuchsin, applying enough heat for gentle steaming.
2. Rinse in tap water.
3. Decolorize in acid alcohol until only a suggestion of pink remains.
4. Wash in tap water.
5. Counterstain with the methylene blue solution for 1 min.
6. Wash in tap water.
7. Air dry and examine under oil.

Results.   Acid-fast organisms, red; others blue.

Analine Blue—Lactophenol Analine Blue

Mounting medium and stain for fungi: Prepare lactophenol and add 0.5% analine blue. (See Lactophenol below.)

Cotton Blue (Methyl Blue)—Lactophenol Cotton Blue

Mounting medium and stain for fungi: Prepare lactophenol and add 0.5% cotton blue (methyl blue). (See Lactophenol below.)

Cytoplasmic Polyhedrosis Virus Detection (Sikorowski et al., 1971)

Staining Solution

The staining solution should be freshly prepared before use.

Buffalo black NBR (Allied Chemical) . . . 0.1 g
100% methyl alcohol . . . . 50.0 ml
Distilled water . . . . 20.0 ml
Glacial acetic acid . . . . 30.0 ml

Preparation of Smear

1. Larvae 4 days old or older are starved for 12 to 24 hr.
2. Place larvae in 100% ethyl alcohol for 5 min.
3. Remove larva from alcohol, place on paper towel and allow to dry.
4. Dissect out the midgut from between the last thoracic legs and the first pair of pro-legs.
5. Mash the midgut section with a flat tooth pick to prepare a thin smear.
6. Air dry smear for 1–2 hr at room temperature.

Staining Procedures

1. Cover smear with stain for 5 min at 40°C (or 20 min at room temperature). Do not allow stain to dry during this period.
2. Drain slide and allow to air dry.
3. Wash gently in tap water for 5 sec.
4. Air dry and examine without cover glass under oil immersion (1000×).

Results. Polyhedra are stained navy blue, while background material is stained light blue.

Feulgen–Schiff Reaction for Granulosis Virus Capsules in Tissue

Reagents

1. *Fixative*—10% formalin in phosphate buffer at pH 6.8–7.0. Fix material at 4.0°C for up to 72 hr.
2. *Schiff's Reagent*—Dissolve 1 g of basic fuchsin in 200 ml

boiling distilled water. Shake for 5 min and cool to 50°C. Filter, and add to the filtrate 20 ml 1 N HCl. Cool to 25°C and add 1 g of sodium or potassium metabisulfite ($Na_2S_2O_5$). Place this solution in the dark for 14–24 hr. Add 2 g activiated charcoal and shake for 1 min. Filter. Keep filtrate in the dark at 0–4°C. Allow to reach room temperature before use.

3. *Acid for Hydrolysis*—1 N hydrochloric acid.
4. *Metabisulfite Solution* (prepare fresh)—Dilute 5 ml 10% aqueous potassium or sodium metabisulfate and 5 ml 1 N HCl in 90 ml distilled water.

Sections.   Dehydrate fixed material through alcohol series to xylol and embed in paraffin or paraplast. Cut sections 5 μm thick and fix to slides.

Method
1. Bring sections to water.
2. Rinse briefly in cold 1 N HCl.
3. Place in 1 N HCl at 60°C (preheated) for hydrolysis for 8 min.
4. Rinse briefly in cold 1 N HCl and then distilled water.
5. Transfer to Schiff's reagent for 30–60 min at room temperature.
6. Drain and rinse in three changes of freshly prepared metabisulfite solution.
7. Rinse in water.
8. Dehydrate in alcohol series, clear in xylol and mount in balsam or synthetic resin.

Results.   Granulosis virus capsules and virogenic stroma appear reddish-purple (violet) in color.

Flagella Stain

Bacterial flagella, often used as taxonomic characters, are very fine organelles of locomotion. In order to be seen under the light microscope, they must be treated in some manner to increase their dimension. The manner described here is taken from Leifson (1960).

Culture. Flagellated bacterial cells are more readily found in young cultures, especially in a phosphate-enriched broth medium. Nutrient broth with 0.1% potassium phosphate added is a good general flagella broth. Culture test organisms 16 hr or less at 20°C in 3.0 ml flagella broth.

Fixation. Add 6.0 ml 10% formlin to the above broth culture.

Wash
1. Dilute the fixed culture with distilled water and centrifuge for 30 min at 3000 RPM in a clinical centrifuge.
2. Discard supernatant, suspend pellet in distilled water, and centrifuge as above.
3. Repeat.
4. Suspend final pellet in distilled water so that it is barely turbid.

Slide Preparation
1. Clean slides overnight in hot (70°–80°C) sulfuric acid saturated with potassium dichromate.
2. Rinse slides thoroughly in tap water, then distilled water, then air dry. Do not touch cleaned slides with anything but clean forceps. They must be kept absolutely grease free. Store in a clean, dry, airtight container.
3. Just before use, heat a slide in the colorless flame of a bunsen burner (the side to be used against the flame), and draw a line with a wax pencil transversely across the slide from side to side about 1/3 the distance from one edge. The slide should be handled only by this edge.
4. A drop of the final bacterial suspension is placed on the distal end of the cooled slide; the slide is tilted to cause the suspension to run down to the wax line. Two such smears, side by side, are readily made on each slide. When the smear has air dried, it is ready to be stained.

Stain
1. Prepare three stock solutions:
   a. 1.2% Basic fuchsin in 95% ethyl alcohol
   b. 3.0% Tannic acid in distilled water
   c. 1.5% Sodium chloride in distilled water
2. Prepare the stain by mixing equal parts of the three stock solutions. The stain solution may be stored for 1 week at room temperature, 1 to 2 months under refrigeration, and indefinitely if frozen.

Stain Application
1. Place the prepared slide on a staining rack, and flood with the staining solution for 5–15 min. (Shorter time for new and warm stain, and longer for old and/or cold stain.)
2. Wash all the stain off the slide at once with running tap water. Do not allow the stain to run off the slide before it is placed under running water.
3. Air dry and examine under oil for flagella.

Giemsa Stain

Giemsa is a commercially prepared blood stain which may be purchased from many biological supply houses. It has many uses in microbiology. For general use in diagnostic work, the stock solution may be diluted at the rate of one drop per milliliter of distilled water or buffer.

Giemsa with HCl Hydrolysis—To Differentiate Rickettsia from Granulosis Virus Capsules

1. Air dry smear.
2. Fix in any suitable fixative, e.g., Carnoy's for 5 min, or methanol, 3–4 min.
3. Hydrolyze in 0.1% HC1 for 2 min.
4. Stain in Giemsa diluted 3 drops stock to 2 ml HOH for 15–45 min.
5. Wash in several changes of HOH.

6. Air dry.
7. Examine under oil, or dehydrate and mount.
8. Results: Rickettsia are strongly colored (positive) in contrast to granulosis virus capsules.

Giemsa Stain for Microsporidan Spores

(Also see Giemsa with HC1 Hydrolysis.)
1. Air dry smear.
2. Fix in methyl alcohol 3–4 min.
3. Air dry, or blot dry.
4. Dilute stock solution 1 drop to 1 ml distilled water.
5. Stain smear for 15 min.
6. Wash in distilled water.
7. Air dry, or blot dry.
8. Examine under oil, or dehydrate and mount.

Giemsa—Slow Giemsa Staining with Acid Hydrolysis for Virus Inclusion Bodies

Buffer.   In 185.0 ml distilled water, mix 1 ml of 0.5 M KH_2PO_4 and 1.5 ml of 0.5 M Na_2HPO_4. Final pH should be 6.89.

Giemsa Stain.   Mix 0.5 ml Giemsa with 50 ml buffer.

Method
1. Fix sections, smear, etc., in Bouin Dubosq Brasil (see Iron Hematoxylin Stain) for 10 min.
2. Wash in 70% ethyl alcohol for three changes, one change every hour.
3. Rinse in distilled water for 5 min.
4. Rinse in Giemsa buffer for 5 min.
5. Stain in buffered Giemsa overnight.
6. Rinse in distilled water.
7. Differentiate in 80% ethyl alcohol with 5% glacial acetic acid (for hydrolysis) until the general blue color of the slide disappears and a spotty blue color remains.
8. Drain and immerse in a 1:1 mixture of acetone–xylene.

9. Pass through two changes of xylene.
10. Mount in balsam, or synthetic resin.

Results. Inclusion bodies are purple in color, and virus rods (if they can be seen) will be red.

Giemsa Stain for Virus Polyhedrosis Inclusions

1. Air dry smear.
2. Treat air dried smear with 0.1% HCl 2 to 5 min.
3. Stain in diluted Giemsa 5 to 10 min.
4. Rinse in running water 5 to 10 sec.
5. Air dry and examine under oil.
6. Results: Polyhedrosis inclusions stain blue and are, therefore, more visible than normal under bright field illumination. Fat globules will stain purple to red, while other crystals, such as ureates, will not stain.

Gram Stain

The method of Gram staining offered here is based on the work of Bartholomew (1962) and is recommended for producing less variable results than other methods.

Dye Formulas

1. *Ammonium Oxalate Crystal Violet* (Hucker modification). *Solution A:* Dissolve 4 g crystal violet (90% dye content) in 40 ml 95% ethyl alcohol. *Solution B:* Dissolve 1.6 g ammonium oxalate in 160 ml distilled water. Mix solutions A and B. It is recommended that the resulting solution be allowed to stand 48 hr before use.

2. *Gram's Iodine.* Place 2 g potassium iodide into a mortar, add 1 g of iodine and grind with a pestle for 5 to 10 sec. Add 1 ml of distilled water and grind. The iodine and potassium iodide should now be in solution. Add 10 ml water and mix. Pour into a reagent bottle and rinse the mortar and pestle with sufficient water to bring the final volume to 200 ml.

3. *Counterstain.* Add 20 ml of 2.5% safranin (86% dye content) in 95% ethyl alcohol to 180 ml distilled water.

Procedure

1. Air dry smear and heat fix by passing lightly through a bunsen flame a few times.
2. Place slide on a staining rack and flood with ammonium oxalate crystal violet for 1 min.
3. Rinse in tap water running in a beaker for 5 sec.
4. Rinse slide with Gram's iodine, then flood slide with this solution for 1 min.
5. Rinse in running tap water 5 sec.
6. Pass the wet slide through three changes of *n*-propyl alcohol in separate coplin jars, 1 min each.
7. Rinse in running tap water 5 sec.
8. Rinse slide with safranin counterstain, then flood with counterstain for 1 min.
9. Rinse in running tap water for 5 sec, then air dry.
10. Examine under oil immersion.

Results.   Gram-positive organisms blue-violet; Gram-negative organisms red. *Caution:* Many organisms are Gram variable, particularly in older cultures. Bacteria used for Gram stain should be from young cultures, preferable 8–16 hr, and no older than 24 hr.

Guegen's Solution—Recommended Mounting Medium and Stain for Fungi

1. Prepare lactophenol and heat (see Lactophenol below).
2. Saturate hot lactophenol with Sudan III, let cool, and filter.
3. Add 0.5% cotton blue (methyl blue), or analine blue.

Iron Hematoxylin Stain for Granulosis Virus Capsules in Tissue (Huger, 1961)

Iron Alum.   2.5% aqueous ferric ammonium sulfate

Heidenhain's Iron Hematoxylin.   Stock solution: Freshly prepared hematoxylin is not good for staining, and must be "ripened." A stock solution may be prepared by dissolving 10% of hematoxylin crystals in 95% ethyl alcohol. It will ripen after several months and may be good for more than a year, but it will not keep indefinitely. *Stain:* 0.5% to 1.0% aqueous hematoxylin. This may be prepared from the above ripened stock, or hematoxylin crystals may be dissolved directly in water and then allowed to stand ("ripen") for 3–6 weeks.

Fixative (Bouin Dubosq Brasil)

80% ethyl alcohol	150 ml
Formalin	60 ml
Glacial acetic acid	15 ml
Picric acid	1 g

Method 1:

a. Fix infected larvae or tissues in Bouin Dubosq Brasil overnight.
b. Wash in 70% ethyl alcohol, dehydrate, clear and embed according to the usual paraffin or paraplast methods.
c. Cut sections 3 to 5 μm in thickness and fix them to slides.
d. Dewax in xylol and pass through descending series of ethyl alcohol to distilled water.
e. Hydrolyze in 50% acetic acid at room temperature for 5 min.
f. Rinse well in distilled water.
g. Mordant in 2.5% iron alum for 2 hr.
h. Rinse thoroughly in distilled water.
i. Stain in Heidenhain's iron hematoxylin for 5 hr.
j. Differentiate carefully in 2.5% iron alum solution with frequent rinsing and microscopic examinations until the black nuclei and the characteristic deep black network of diseased cells become clearly outlined.
k. Wash in running tap water for 1 hr, and rinse in distilled water for 10 min.
l. Counterstain in 0.5% aqueous erythrosin for 2 min.

m. Dehydrate in ascending series of ethyl alcohol (70%, 80%, 95%, 100%), passing quickly to 95%, clear in xylol, and mount in balsam or synthetic resin.

Results.   The capsules are selectively stained in a beautiful intense bright red color, and even single ones phagocytosed by blood cells are stained. At first sight they are clearly contrasted from the pale gray violet colored cytoplasm. Moreover, the typical network mentioned above is stained deep black, similar to the nuclei.

Method 2

a–d. Same as Method 1.

e.   Hydrolyze in 1 N HCl for 10–20 min at 60°C.

f.   Rinse well in distilled water.

g.   Mordant in 2.5% iron alum for 5–10 hr.

h.   Rinse thoroughly in distilled water.

i.   Stain in Heidenhain's iron hematoxylin for 15–20 hr.

j.   Differentiate carefully in 2.5% iron alum with frequent rinsing and microscopic examinations until the excess coloration is removed so that the capsules appear blue-black to black and structural details of the black network become visible.

k–m. Same as Method 1.

Results.   The capsules show a striking blue-black color. If the sections are thin enough, the individual capsules become sharply outlined; otherwise they appear as blue-black to black masses. The network of infected cells is stained deep black. The nuclei appear black if hydrolysis was 10–15 min, and red if hydrolysis was 20 min or more. The cytoplasm is colored pale red violet (10-min hydrolysis) to red (20 min hydrolysis). This method is also convenient for staining smears, with hydrolysis extended to about 30 min in order to obtain deep black colored capsules.

Lactophenol—Mounting Medium for Fungi, and Base for Cotton Blue, Analine Blue, and Guegen's Stains

Phenol crystals .... 100 g
Lactic acid (USP 85%) .... 80 ml
Glycerine .... 159 ml
Distilled water .... 100 ml

Macchiavello Stain

1. Air dry smear.
2. Fix 5–10 min in methanol.
3. Stain 60 min in 1% basic fuchsin.
4. Rinse in HOH.
5. Differentiate 5 sec in 0.5% citric acid.
6. Rinse in HOH.
7. Counterstain for 20 sec in 1% methylene blue.
8. Rinse in HOH.
9. Air dry and examine under oil, or dehydrate to xylene and mount in permount.
10. Results: Rickettsia, crystals, and globules of spheroidocytes, red. Albuminoid crystals and bacteria, blue.

Neutral Red Stain

1. Air dry smear.
2. Stain 10–15 min in 0.5–1.0% neutral red in 100% ETOH.
3. Rinse.
4. Air dry.
5. Examine under oil, or dehydrate and mount.
6. Results: NR bodies in rickettsial infections stain red.

Sudan III—To Differentiate Virus Polyhedra from Fat Droplets

1. Air dry smear.
2. Stain 10–15 min in saturated aqueous Sudan III.

3. Rinse 5–10 sec in running tap water.
4. Air dry and examine under oil.
5. Results: Fat droplets stain red while polyhedra remain unstained.

MISCELLANEOUS TECHNIQUES

Bacteriological Loop (Figure 105)

Wire inoculating or transfer loops are used for making streak plates, inoculating broth cultures, transferring inocula to slides for microscopic examination, etc. The loop is made from 24–26 gauge

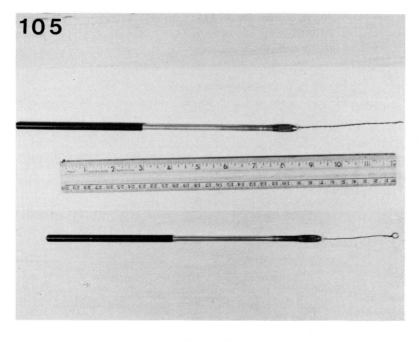

FIGURE 105

Bacteriological loop for streak plate cultures, and twisted wire for stab cultures.

platinum or nichrome wire fixed to a standard bacteriological wire-loop holder. The wire is formed into a loop about 3 mm in diameter with a shaft about 8 cm long.

Bacteriological Needle or Wire for Stab Cultures

A needle about 8 cm long may be used for stab cultures, or a suitable substitute may be made from loop wire by twisting a double length of wire back on itself and leaving the last 15 mm single for insertion into the holder (Figure 105).

Galleria mellonella—Rearing Method

Rearing Medium
1. Mix together 100 ml water, 100 ml honey, 100 ml glycerine, and 5 ml Deca Vi-Sol or equivalent vitamins.
2. Pour the liquid mixture into 1200 cm³ dry Pablum or Gerber's mixed cereal.
3. Mix until homogeneous and place in a one-half gallon mason jar with a screen top. (A circle may be cut from a regular top and a piece of ordinary window screen cut to fit.)
4. Place 200–300 freshly collected eggs on the medium and incubate at 30°C, or at room temperature (30°C is preferable).

The larvae will reach the last instar in 4–5 weeks at 30°C, and take about a week or more at room temperature. The culture may be stored at 10°C for up to 3 months, after which it should be returned to rearing temperature and the remaining larvae allowed to pupate. Emerging adults are collected by anesthetizing with CO_2 or chilling in a refrigerator and are transferred to a clean gallon mason jar with a screen top. New adults are added to the jar as they emerge. It is not necessary to provide the adults with food or water. A piece of accordion-pleated wax paper provides an oviposition site and unfolding the wax paper causes the eggs to fall off.

Koch's Postulates

The following postulates, devised by the German bacteriologist Robert Koch (1843–1910), provide a general outline which may be used when conducting infectivity tests. If these steps are carried out with positive results, it may be considered conclusive evidence that the microorganism in question is the cause of the disease.

1. The microorganism must be present in every case of the disease.
2. The microorganism must be isolated in pure culture.
3. The microorganism in pure culture, when inoculated into a susceptible animal, must give rise to the disease.
4. The same microorganism must be present in, and recoverable from, the experimentally diseased animal.

Microscopic Preparations—Wet Mount and Tissue Smear

Wet Mount.   A wet mount is simply made by placing a bit of tissue or hemolymph in a drop of water or Ringer's solution on a microscope slide, and covering it with a cover glass. The preparation can be kept longer by ringing the cover glass with *Vaspar* or nail polish.

Vaspar.   Vaspar is prepared by melting together equal parts of Vaseline (petroleum jelly) and paraffin. It may be kept in a small crucible or other container and melted over a bunsen flame when needed. It is applied with a small watercolor brush in a thin line to seal the edges of a cover glass.

Tissue Smear.   Tissue smears are prepared by dissecting out a small amount of tissue which is squashed and spread on a microscope slide and allowed to air dry. The thinner the preparation the better it will be for microscopic examination. A wet mount may be used for this purpose by removing the cover slip and allowing it to air dry. Smears may be fixed by gentle heating, or by chemical fixation as prescribed in the staining procedure.

Per Os Injection

Viruses, rickettsia, bacteria, protozoa, and some fungi normally enter their insect host through the mouth (per orally) and penetrate the gut wall into the hemocoel. This route of infection is termed *per oral,* or *per os,* and when testing such microorganisms for pathogenicity, a method for per os inoculation should be used.

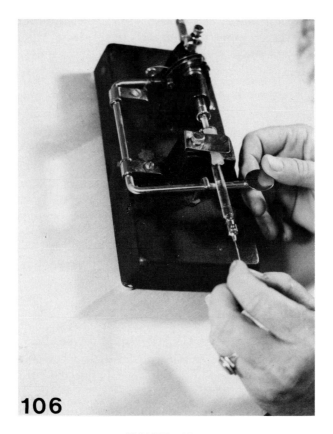

106

FIGURE 106

The Dutky–Fest microinjector showing a hypodermic syringe fitted with a glass-tipped needle used for per os injections.

The simplest method is to contaminate the insect's food with a suspension of the suspected pathogen. A more precise method is to force feed the inoculum to an insect by per os injection. One way this can be done is with a Dutky–Fest microinjector equipped with a 1-cm³ tuberculin syringe and a fine glass-tipped needle (Figure 106).

Glass-Tipped Feeding Needle.   The feeding needle is prepared by drawing out a fine capillary tube so it will fit into the mouth of the test insect. Cut the capillary at the small end and round this in a light flame, being careful not to seal the opening. The large end is sealed to the base of a 27-gauge needle with a drop of Duco cement.

Method.   The pathogen suspension is drawn into the sterile syringe fitted with a feeding needle. The syringe is mounted on the microinjector and the delivery calculated. The tip of the needle is teased into the mouth of a test insect and the desired amount of inoculum injected. Care must be taken not to puncture the esophagus, as this will allow normal gut bacteria to enter the hemocoel and cause a fatal septicemia. After injection the insect is reared on normal food. Several insects should be used for each test, and a control should be run using sterile saline, or sterile distilled water.

Stab Cultures

Certain bacteriological media such as nutrient gelatin and triple sugar iron agar require inoculation by the stab method. This is accomplished by taking a wire inoculating needle, flaming it red hot, cooling it in alcohol, burning off the alcohol, dipping the needle in the inoculum, and inserting it deeply into the test medium.

Streak Plate Method (Figure 107)

The streak plate method is used to obtain separate colonies of a single bacterial species from a mixed culture, or from a specimen containing a mixed flora, or to subculture pure isolates from one

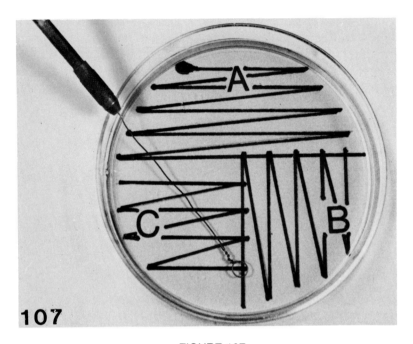

FIGURE 107

Illustration of the procedure described for the streak plate method.

culture to another. Suspensions used for streak plates should be barely turbid.

Materials Needed
   Petri plate with solid agar medium
   Bacteriological inoculating loop (Figure 105)
   Suspension of inoculum
   Bunsen burner, or alcohol burner

Procedures.   Flame the loop along its entire length until red hot, then quickly pass the holder through the flame for a few inches. Cool the loop by dipping it into 70% alcohol and flaming off the excess.

1. Place a loopful of inoculum close to the edge of the agar in a petri dish, and spread on area A as shown in Figure 107. In

doing this, raise the cover of the petri dish at one side just high enough to allow manipulation of the inoculating loop.

2. Flame the loop, and turning the plate 90 degrees, cover area B as illustrated, passing over the streak in A each time.

3. Turn the plate again and streak over area C without flaming the loop.

4. Incubate the plate in the inverted position. Isolated colonies should be found in area C if the inoculum was heavy, and in area B and possibly area A if the inoculum was lighter.

Virus Inclusion Bodies—Dissolution in Alkalai

The dissolution of virus inclusion bodies in alkalai may be demonstrated as follows: Prepare a wet mount of the material under investigation and place a drop of 1 N NaOH at the edge of the cover slip, allowing it to enter while observing under high dry (400×) magnification. As the alkalai flows through, most virus inclusion bodies will swell and dissolve. The swelling is difficult to see in granulosis capsules, but is quite evident with polyhedrosis inclusions (Figure 58).

LITERATURE CITED

Afrikian, E. K. 1973. *Entomopathogenic Bacteria and Their Significance.* Akad. Nauk. Armianskoi SSR. Erevan. 418 pp. (In Russian).

Ainsworth, G. C. 1961. *Ainsworth & Bisby's Dictionary of the Fungi,* 5th Ed. Commonwealth Mycological Institute, Kew, Surrey, England. 547 pp.

Ainsworth, G. C., F. K. Sparrow, and A. S. Sussman. 1973. *The Fungi, an Advanced Treatise,* Vol. IVA (621 pp.) and IVB (504 pp.). Academic Press, New York.

Aruga, H., and Y. Tanada (Eds). 1971. The cytoplasmic-polyhedrosis viruses of the silkworm. Univ. Tokoyo Press, Tokoyo. 234 pp.

Bailey, L., A. J. Gibbs, and R. D. Woods. 1964. Sacbrood virus of the larval honeybee (*Apis mellifera* Linnaeus). *Virology* **23:**425–429.

Bailey, L. 1973a. Control of invertebrates by viruses. In: *Viruses and Invertebrates* (Ed. A. J. Gibbs). North-Holland, Amsterdam. 533–553.

Bailey, L., 1973b. Viruses and hymenoptera. In: *Viruses and Invertebrates* (Ed. A. J. Gibbs). North-Holland, Amsterdam. 442–454.

de Barjac, H., and A. Bonnefoi. 1973. Mise au point sur la classification des *Bacillus thuringiensis. Entomophaga* **18:**5–17.

Barnett, H. L., and B. B. Hunter. 1972. *Illustrated Genera of Imperfect Fungi,* 3rd. Ed. Burgess Publishing Co., Minneapolis, Minn., 241 pp.

Bartholomew, J. W. 1962. Variables influencing results and precise definition of steps in gram staining as a means of standardizing the results obtained. *Stain Technology* **37:**139–155.

Baudoin, J. 1969. Nouvelles espèces de Microsporidies chez des larves de Trichopteres. *Protistologica* **5:**441–446.

Bell, J. V. 1974. Mycoses. In: *Insect Diseases,* Vol. I. (Ed. G. E. Cantwell). Marcel Dekker, Inc., New York. 185–236.

Bell, J. V., and R. J. Hamalle. 1974. Viability and pathogenicity of entomogenous fungi after prolonged storage on silica gel at −20°C. *Can. J. Microbiol.* **20:**639–642.

Bergoin, M., and S. Dales. 1971. Comparative observations on poxviruses of invertebrates and vertebrates. In: *Comparative Virology* (Eds. K. Maramorosch and E. Kurstak). Academic Press, N. Y. 171–205.

Brooks, M. A. 1974. Genus VIII *Symbiotes* and Genus IX *Blattabacterium*. In: *Bergey's Manual of Determinative Bacteriology* (Eds. R. E. Buchanan and N. E. Gibbons). Williams & Wilkins Co., Baltimore. 900–901.

Brooks, W. M. 1974. Protozoan infections. In: *Insect Diseases,* Vol. 1. (Ed. G. E. Cantwell). Marcel Dekker, Inc., New York. 237–300.

Brown, A.H.S., and G. Smith. 1957. The genus *Paecilomyces* Bainier and its perfect stage *Byssochlamys* Westling. *Trans. British Mycological Soc.* **40:**17–89.

Buchanan, R. E., and N. E. Gibbons (Eds.). 1974. *Bergey's Manual of Determinative Bacteriology,* 8th Ed. Williams & Wilkins Co., Baltimore. 1268 pp.

Bucher, G. E. 1959. Bacteria of grasshoppers of western Canada. III. Frequency of occurrence, pathogenicity. *J. Insect Pathol.* **1:**391–405.

Bucher, G. E. 1961. Control of the eastern tent caterpillar, *Malacosoma americanum* (Fabricius), by distribution of spores of two species of *Clostridium*. *J. Insect Pathol.* **3:**439–445.

Bucher, G. E. 1963. Nonsporulating bacterial pathogens. In: *Insect Pathology, an Advanced Treatise,* Vol. II (Ed. E. A. Steinhaus). Academic Press, New York. 117–147.

Burges, H. D., and N. W. Hussey. 1971. *Microbial Control of Insects and Mites.* Academic Press, New York and London. 861 pp.

Burkholder, W. E., and R. J. Dicke. 1964. Detection by ultraviolet light of stored product insects infected with *Mattesia dispora*. *J. Econ. Entomol.* **57:**878–879.

Canning, E. U. 1956. A new eugregraine of locusts, *Gregarina garnhami* n.sp. parasitic in *Schistocerca gregaria* Forsk. *J. Protozool.* **3:**50–62.

Canning, E. U. 1957. On the occurrence of *Plistophora culicis* Weiser in *Anopheles gambiae*. *Rivista de Malariologia* **36:**39–50.

Cantwell, G. E. 1974. Honey bee diseases, parasites, and pests. In: *Insect Diseases* (Ed. G. E. Cantwell). Vol. II. Marcel Dekker, Inc., New York. 501–547.

Caullery, M., and F. Mesnil. 1905. Recherches sur les Haplosporidies. *Arch. Zool. Exp. Gén.* **4:**101–180.

Clark, T. B., and T. Fukuda. 1971. *Plistophora chapmani* n. sp. in *Culex territans* from Louisiana. *J. Invertebr. Pathol.* **18:**400–404.

Clark, T. B., W. R. Kellen, J. E. Lindegren, and R. D. Sanders. 1966. *Pythium* sp. (Phycomycetes:Pythiales) pathogenic to mosquito larvae. *J. Invertebr. Pathol.* **8:**351–354.

Codreanu, M. 1940. Sur quatre grégarines nouvelles du genre *Enterocystis,* parasites des éphémerès torrenticoles. *Arch. Zool. Exp. Gén.* **81:**113–122.

Codreanu, R. 1963. On the structure of the spore of *Telomyxa glugeiformis* Léger et Hesse 1910 and the general classification of microsporidia. In: *Progress in Protozoology*. (Eds J. Zudvik, J. Lom, and J. Vavra). Academic Press, New York. 82–84.

Codreanu, R., and S. Vavra. 1970. The structure and ultrastructure of the microsporidan *Telomyxa glugeiformis* Léger and Hesse, 1910 parasite of *Ephemera danica* (Mull) nymphs. *J. Protozool.* **17**:374–384.

Corliss, T. O. 1960. *Tetrahymena chironomini* sp. nov., a ciliate from midge larvae, and the current status of facultative parasitism in the genus *Tetrahymena. Parasitology* **50**:111–153.

Couch, J. N. 1937. A new fungus intermediate between the rusts and *Septobasidium. Mycologia* **29**:665–673.

Couch, J. N. 1938. *The Genus Septobasidium.* The University of North Carolina Press, Chapel Hill, N.C. 480 pp.

Couch, J. N., and C. J. Umphlett. 1963. *Coelomomyces* infections. In: *Insect Pathology, an Advanced Treatise,* Vol. II (Ed. E. A. Steinhaus). Academic Press, New York. 149–188.

David, W.A.L. 1975. The status of viruses pathogenic for insects and mites. *Ann. Rev. Entomol.* **20**:97–117.

Delgarno, L., and M. W. Davey. 1973. Virus replication. In: *Viruses and Invertebrates* (Ed. A. J. Gibbs). North-Holland, Amsterdam. 245–270.

DeHoog, G. S. 1972. The genera *Beauveria, Isaria, Tritirachium* and *Acrodontium* gen. nov. Centraalbureau voor Schimmelcultures, Baarn. *Studies in Mycology* **1**:1–41.

Difco Laboratories. 1953. *Difco Manual of Dehydrated Media and Reagents for Microbiological and Clinical Laboratory Procedures.* 9th Edition. Difco Laboratories, Inc., Detroit, Michigan. 350 pp.

Dissanaike, A. S. 1955. A new Schizogregarine *Triboliocystis garnhami* n.g., n.sp., and a new microsporidian *Nosema buckleyi* n. sp. from the fat body of the flour bettle *Tribolum castaneum. J. Protozool.* **2**:150–156.

Doane, C. C., and J. J. Redys. 1970. Characteristics of motile strains of *Streptococcus faecalis* pathogenic to larvae of the gypsy moth. *J. Invertebr. Pathol.* **15**:420–430.

Doby, J., and F. Saguez. 1964. *Weiseria,* genre nouveau de Microsporidies et *Weiseria laurenti* n. sp., parasites de larves de *Prosimulium inflatum* Davies, 1957 (Diptères Paranématocères). *Compt. Rend. Acad. Sci. Paris.* **259**:3614–4617.

Dutky, S. R. 1963. The milky diseases. In: *Inseçt Pathology, an Advanced Treatise,* Vol. II (Ed. E. A. Steinhaus). Academic Press, New York. 75–115.

Evlakhova, A. A. 1974. *Entomogenous Fungi, Classification, Biology and Practical Significance.* Nauka, Leningrad. 260 pp. (In Russian)

Falcon, L. A. 1971. Use of bacteria for microbial control. In: *Microbial Control of Insects and Mites* (Eds. H. D. Burges and N. W. Hussey). Academic Press, New York. 67–95.

Faust, R. M. 1974. Bacterial diseases. In: *Insect Diseases,* Vol. I (Ed. G. E. Cantwell). Marcel Dekker, Inc. New York. 87–183.

Ganhão, J.F.P. 1956. *Cephalosporium lecanii* Zimm. um fungo entomogeno de cochonilhas. *Broteria* **25**:71–135.

Glinski, Z. 1968. Bacteriological diagnostics of the bee foul brood. *Medy. Veter.* **24:**346–349.

Greenberg, B. 1971. *Flies and Disease.* Vol. I. *Ecology, Classification and Biotic Associations.* Princeton Univ. Press, Princeton, New Jersey. 856 pp.

Harrap, K. A. 1973. Virus infection in invertebrates. In: *Viruses and Invertebrates* (Ed. A. J. Gibbs). North-Holland, Amsterdam. 271–299.

Hassan, S., and C. Vago. 1972. The pathogenicity of *Fusarium oxysporum* to mosquito larvae. *J. Invertebr. Pathol.* **20:**268–271.

Hazard, E. I., and D. W. Anthony. 1974. A redescription of the genus *Parathelohania* Codreanu 1966 (Microsporida: Protozoa) with a reexamination of previously described species of *Thelohania* Henneguy 1892 and descriptions of two new species of *Parathelohania* from anopheline mosquitoes. U.S. Dept. Agr. Tech. Bull. No. 1505. 26 pp.

Hazard, E. I., and K. E. Savage. 1970. *Stempellia lunata* sp. n. in larvae of the mosquito *Culex pilosus* collected in Florida. *J. Invertebr. Pathol.* **15:**49–54.

Hazard, E. I., and J. Weiser. 1968. Spores of *Thelohania* in adult female *Anopheles:* Development and transovarial transmission and redescriptions of *T. legeri* Hesse and *T. obesa* Kudo. *J. Protozool.* **15:**817–823.

Heimpel, A. M. 1967. A taxonomic key proposed for the species of the "Crystalliferous Bacteria." *J. Invertebr. Pathol.* **9:**364–375.

Heimpel, A. M., and T. A. Angus. 1963. Diseases caused by certain spore-forming bacteria. In: *Insect Pathology, an Advanced Treatise,* Vol. II. (Ed. E. A. Steinhaus). Academic Press, New York. 21–73.

Hesse, E. 1935. Sur quelques microsporidies parasites de *Megacyclops viridis. Arch. Zool. Exp. Gén.* **75:**651–661.

Hölldobler, K. 1930. Uber eine merkwürdige Parasitenerkrankung von *Solenopsis fugax. Z. Parasitenk.* **2:**67–72.

Honigberg, B. (Chairman) *et al.* 1964. A revised classification of the phylum Protozoa. *J. Protozool.* **11:**7–20.

Huger, A. 1961. Methods for staining capsular virus inclusion bodies typical of granuloses of insects. *J. Insect Pathol.* **3:**338–341.

Hunter, B. F. 1970. Ecology of waterfowl botulism toxin production. *Transactions of the 35th North American Wildlife Natural Resources Conference.* 9 pp.

Ignoffo, C. M. 1968. Viruses-living insecticides. In: *Insect Viruses.* (Ed. K. Maramorosch). *Curr. Topics Microbiol. Immunol.* **42:**129–167.

Ignoffo, C. M. 1974. Microbial control of insects: Viral pathogens. In: *Proceedings of the Summer Institute on Biological Control of Plant Insects and Diseases* (Eds. F. G. Maxwell and F. A. Harris). Univ. Press, Jackson, Miss. 541–557.

Ishihara, R. 1967. Stimuli causing extrusion of polar filaments of *Glugea fumiferanae* spores. *Can. J. Microbiol.* **13:**1321–1332.

Jahn, T. L. 1949. *How to Know the Protozoa.* Wm. C. Brown Co., Dubuque, Iowa. 234 pp.

Jamnback, H. A. 1970. *Caudospora* and *Weiseria,* two genera of Microsporidia parasitic in blackflies. *J. Invertebr. Pathol.* **16**:3-13.

Kamburov, S. S., D. J. Nadel, and R. Kenneth. 1967. Observations on *Hesperomyces virescens* Thaxter (Laboulbeniales), a fungus associated with premature mortality of *Chiolocorus bipustulatus* L. in Israel. *Israel J. Agric. Res.* **17**:131-134.

Karling, J. S. 1948. Chytridiosis of scale insects. *Am. J. Botany* **35**:246-254.

Keilin, D. 1920. On two new gregarines, *Allantocystis dasyhelei* n.g., n. sp., and *Dendrohyrhynchus systeni* n.g., n. sp. parasitic in the alimentary canal of the dipterous larvae, *Dasyhelea obscura* Winn. and *Systenus* sp. *Parasitology* **12**:154-158.

Kellen, W. R., T. B. Clark, J. E. Lindegren, B. C. Hoe, M. H. Rogoff, and S. Singer. 1965. *Bacillus sphaericus* Neide as a pathogen of mosquitoes. *J. Invertebr. Pathol.* **7**:442-448.

Kellen, W. R., and J. E. Lindegren. 1974. Life cycle of *Helicosporidium parasiticum* in the navel orangeworm, *Paramyelois transitella. J. Invertebr. Pathol.* **23**:202-208.

Kellen, W. R., J. E. Lindegren, and D. F. Hoffmann. 1972. Developmental stages and structure of a *Rickettsiella* in the naval orangeworm, *Paramyelois transitella* (Lepidoptera: Phycitidae). *J. Invertebr. Pathol.* **20**:193-199.

Koval, E. Z. 1974. *Guidebook to Entomophilic Fungi of the USSR.* Naukova Dumka, Kiev. 260 pp. (In Russian)

Kramer, J. P. 1959. Studies on the morphology and life history of *Perezia pyraustae* Paillot (Microsporidia: Nosematidae). *Trans. Am. Microsc. Soc.* **78**:336-342.

Kramer, J. P. 1964. *Nosema kingi* sp. n., a microsporidian from *Drosophila willestoni* Sturtevant, and its infectivity for other muscoids. *J. Insect Pathol.* **6**:491-499.

Krieg, A. 1963. Rickettsiae and Rickettsioses. In: *Insect Pathology, an Advanced Treatise,* Vol. I (Ed. E. A. Steinhaus). Academic Press, New York. 577-617.

Krieg, A. 1971. Possible use of Rickettsiae for microbial control of insects. In: *Microbial Control of Insects and Mites* (Eds. H. D. Burges and N. W. Hussey). Academic Press, New York. 173-179.

Kudo, R. 1924. A biologic and taxonomic study of the Microsporidia. *Ill. Biol. Monogr.* **9**:1-268.

Kudo, R. 1942. On the microsporidian, *Duboscqia legeri,* Perez 1908, parasitic in *Reticulitermes flavipes. J. Morphol.* **71**:307-333.

Kudo, R. 1966. *Protozoology.* 5th Ed. Charles C Thomas, Springfield, Ill. 1174 pp.

Leifson, E. 1960. *Atlas of Bacterial Flagellation.* Academic Press, New York. 171 pp.

Léger, L. 1926. Sur *Trichoduboscqia epori* Léger, Microsporidie parasite des larves d'Ephémerides. *Travaux du Lab. d'Hydrobiologie et des Pisciculture de l'Universite de Grenoble.* **18**:9-14.

Léger, L., and E. Hesse. 1905. Sur un nouveau protiste parasite des Otiorrhynques. *Compt. Rend. Soc. Biol.* **58**:92–94.

Lipa, J. J. 1963. Infections caused by protozoa other than sporozoa. In: *Insect Pathology, an Advanced Treatise,* Vol. II (Ed. E. A. Steinhaus). Academic Press, New York. 348–351.

Longworth, J. F. 1973. Viruses and Lepidoptera. In: *Viruses and Invertebrates* (Ed. A. J. Gibbs). North-Holland, Amsterdam. 428–441.

Ludwig, F. W. 1947. Studies on the protozoan fauna of the larvae of the crane fly *Tipula abdominalis.* II. The life history of *Ithania wenrichi* n. gen., n. sp., a coccidian found in the caeca and mid-gut, and a diagnosis of Ithaniinae n. subfamily. *Trans. Am. Microsc. Soc.* **66**:22–33.

Lysenko, O. 1963. The taxonomy of entomogenous bacteria. In: *Insect Pathology, an Advanced Treatise,* Vol. II (Ed. E. A. Steinhaus). Academic Press, New York. 1–20.

Machado, A. 1913. Zytologie und Entwicklungszyklus der *Chagasella alydi,* einer neuen Kokzidienart aus einer Wanze vom "Genus *Alydus.*" *Mem. Inst. Oswaldo Cruz* **5**:32–44.

MacLeod, D. M. 1963. Entomophthorales infections. In: *Insect Pathology, an Advanced Treatise* (Ed. E. A. Steinhaus). Academic Press, New York. 189–231.

Madelin, M. F. 1963. Diseases caused by Hyphomycetous fungi. In: *Insect Pathology, an Advanced Treatise* (Ed. E. A. Steinhaus). Academic Press, New York. 233–271.

Madelin, M. F. 1966. Fungal parasites of insects. *Ann. Rev. Entomol.* **11**:423–448.

Mains, E. B. 1950. Entomogenous species of *Akanthomyces, Hymenostilbe* and *Insecticola* in North America. *Mycologia* **42**:566–589.

Mains, E. B. 1951. Entomogenous species of *Hirsutella, Tilachlidium* and *Synnematium. Mycologia* **43**:691–718.

Mains, E. B. 1959. North American species of *Aschersonia* parasitic on Aleyrodidae. *J. Insect. Pathol.* **1**:43–47.

Maramorosch, K. 1968a. Plant pathogenic viruses in insects. In: *Insect Viruses* (Ed. K. Maramorosch). *Curr. Top. Microbiol. Immunol.* **42**:94–107.

Maramorosch, K. (Ed.). 1968b. *Insect Viruses. Curr. Top. Microbiol. Immunol.* **42**:192 pp.

Marshall, I. D. 1973. Viruses and Diptera. In: *Viruses and Invertebrates.* (Ed. A. J. Gibbs). North-Holland, Amsterdam. 406–427.

Martignoni, M. E., P. J. Iwai, and L. J. Wickerham. 1969. A candidiasis in larvae of the Douglas-fir tussock moth, *Hemerocampa pseudotsugata. J. Invertebr. Pathol.* **14**:108–110.

Martignoni, M. E., and P. J. Iwai. 1975. A catalog of viral diseases of insects and mites. *USDA Forest Service Gen. Tech. Rpt.* **40**:23 pp.

McEwen, F. L. 1963. *Cordyceps* infections. In: *Insect Pathology, an Advanced Treatise* (Ed. E. A. Steinhaus). Academic Press, New York. 273–290.

McLaughlin, R. E. 1965. *Mattesia grandis* n. sp. a sporozoan pathogen of the boll weevil, *Anthonomis grandis* Boheman. *J. Protozool.* **12**:405–413.

McLaughlin, R. E. 1969. *Glugea gasti* sp. n., a microsporidian pathogen of the boll weevil, *Anthonomus grandis*. *J. Protozool.* **16**:84–92.

McLaughlin, R. E. 1971. Use of protozoans for microbial control of insects. In: *Microbial Control of Insects and Mites* (Eds. H. D. Burges and N. W. Hussey). Academic Press, New York. 151–172.

McLaughlin, R. E. 1973. Protozoa as microbial control agents. *Misc. Pub. Entomol. Soc. Am.* **9**:95–98.

Meade, G. C. 1963. A medium for the isolation of *Streptococcus faecalis*, sensu strictu. *Nature* **197**:1323–1324.

Miller, J. H. 1940. The genus *Myriangium* in North America. *Mycologia* **32**:587–600.

Miller, M. W., and N. van Uden. 1970. In: *The Yeasts* (Ed. J. Lodder). North Holland, Amsterdam. 408–429.

Mims, C. A., M. F. Day, and I. D. Marshall. 1966. Cytopathic effect of Semliki Forest virus in the mosquito, *Aedes aegypti*. *Am. J. Trop. Med. Hyg.* **15**:775–784.

Morrill, A. W., and E. A. Black. 1912. Natural control of white flies in Florida. *USDA Bur. Entom. Bull.* **102**:78 pp.

Moulder, J. W., 1974. Order I Rickettsiales. In: *Bergey's Manual of Determinative Bacteriology*, 8th Ed. (Eds. R. E. Buchanan and N. E. Gibbons). Williams & Wilkins Co., Baltimore. 882–928.

Nenninger, U. 1948. Die Peritrichen der Umgebung von Erlangen mit besonderer Berücksichtigung ihrer Wirtsspezifität. *Zool. Jahrb. Syst.* **77**:169–266.

Niklas, O. F. 1957. Zur Temperaturabhängigkeit der Vertikalbewegungen Rickettsiosekranker Maikafer-Engerlinge (*Melolontha* spec.). *Anz. Schadlingskunde* **30**:113–116.

Noland, L. E., and H. E. Finley. 1931. Studies on the taxonomy of the genus *Vorticella*. *Trans. Am. Microsc. Soc.* **50**:81.

Ormières, R., and V. Sprague. 1973. A new family, new genus, and new species allied to the Microsporida. *J. Invertebr. Pathol.* **21**:224–240.

Page, L. A. 1974. Order II, Chlamydiales. In: *Bergey's Manual of Determinative Bacteriology*, 8th Ed. (Eds. R. E. Buchanan and N. E. Gibbons). Williams & Wilkins Co., Baltimore. 914–918.

Petch, T. 1921. Fungi parasitic on scale insects. Presidential Address. *British Mycological Soc.* **7**:18–24.

Poinar, Jr., G. O. 1975. *Entomogenous Nematodes*. E. J. Brill, Leiden. 317pp.

Poinar, Jr., G. O. 1977. *CIH Key to the Groups and Genera of Nematode Parasites of Invertebrates* (Ed. S. Willmott). Commonwealth Agricultural Bureaux. 43 pp.

Poinar, Jr., G. O., and Thomas, G. 1967. The nature of *Achromobacter nematophilus* as an insect pathogen. *J. Invertebr. Pathol.* **9:**510–514.

Poisson, R. 1941. Les microsporidies parasites des insectes hemipteres. IV. *Arch. Zool. Exp. Gen.* **82:**30–35.

Prasertphon, S., and Y. Tanada. 1968. The formation and circulation, in *Galleria* of hyphal bodies of entomophthoraceous fungi. *J. Invertebr. Pathol.* **11:**260–280.

Rioux, J. A., and F. Achard. 1956. Entomophytose mortelle a *Saprolegnia diclina* Humphrey 1892 dans un élevage d'*Aedes berlandi* Seguy 1921. *Vie et Milieu* **7:**326–337.

Ristic, M., and J. P. Krier. 1974. Family III Anaplasmataceae. In: *Bergey's Manual of Determinative Bacteriology.* 8th Ed. (Eds. R. E. Buchanan and N. E. Gibbons). Williams & Wilkins Co., Baltimore. 906–914.

Roberts, D. W., and W. G. Yendol. 1971. The use of fungi for microbial control of insects. In: *Microbial Control of Insects and Mites* (Eds. H. D. Burges and N. W. Hussey). Academic Press, New York. 125–149.

Sanders, R. D., and G. O. Poinar, Jr. 1973. Fine structure and life cycle of *Lankesteria clarki* sp. n. (Sporozoa: Eugregarinida) parasitic in the mosquito *Aedes sierrensis* (Ludlow). *J. Protozool.* **20:**594–602.

Schaefer, E. 1961. Application of the cytochrome oxidase reaction to the detection of *Pseudomonas aeruginosa* in mixed cultures. *Roentgen Univ. Lab. Praxis* **14:**142–146.

Sen, S. K., M. S. Jolly, and T. R. Jammy. 1970. A mycosis in the Indian tasar silkworm, *Antheraea mylitta* Drury, caused by *Penicillum citrinum* Thom. *J. Invertebr. Pathol.* **15:**1–5.

Sikorowski, P., J. R. Broome, and G. L. Andrews. 1971. Simple methods for detection of cytoplasmic polyhedrosis virus in *Heliothis virescens.* *J. Invertebr. Pathol.* **17:**451–452.

Sinka, R. C. 1973. Viruses and leafhoppers. In: *Viruses and Invertebrates* (Ed. A. J. Gibbs). North-Holland, Amsterdam. 493–511.

Skou, J. P. 1972. Ascosphaerales. *Friesia* **10:**1–24.

Smith, K. M. 1967. *Insect Virology.* Academic Press, New York. 256 pp.

Smith, K. M. 1971. The viruses causing the polyhedroses and granuloses of insects. In: *Comparative Virology* (Eds. K. Maramorosch and E. Kurstak). Academic Press, New York. 479–507.

Smirnoff, W. A. 1974. Réduction de viabilité et de la fécondité de *Neodiprion swainei* (Hymenoptères: Tenthredinidae) par le flagellé *Herpetomonas swainei* sp. n. (Protozoaires). *Phytoprotection* **55:**64–66.

Speare, A. T. 1921. *Massospora cicadina* Peck, a fungus parasite of the periodical cicada. *Mycologia* **13:**72–82.

Sprague, V. 1940. Observations on *Coelosporidium periplanetae* with special reference to the development of the spore. *Trans. Am. Microsc. Soc.* **59:**460–474.

Sprague, V. 1963. Revision of genus *Haplosporidium* and restoration of genus *Minchinia* (Haplosporida, Haplosporidiidae). *J. Protozool.* **10:**263–266.

Sprague, V., R. Ormières, and J. F. Manier. 1972. Creation of a new genus and a new family in the Microsporida. *J. Invertebr. Pathol.* **20**:228-231.

Stairs, G. G. 1971. Use of viruses for microbial control of insects. In: *Microbial Control of Insects and Mites* (Eds. H. D. Burges and N. W. Hussey). Academic Press, New York. 97-124.

Steinhaus, E. A. 1949. *Principles of Insect Pathology.* McGraw-Hill, New York. 757 pp.

Steinhaus, E. A. 1951. Report on diagnoses of diseased insects 1944-1950. *Hilgardia* **20**:629-678.

Steinhaus, E. A. 1959. *Serratia marcescens* Bizio as an insect pathogen. *Hilgardia* **28**:351-380.

Steinhaus, E. A. (Ed.). 1963. *Insect Pathology, an Advanced Treatise.* Academic Press, New York. Vol. I, 661 pp.; Vol. II, 689 pp.

Steinhaus, E. A., and G. A. Marsh. 1962. Report of diagnoses of diseased insects 1951-1961. *Hilgardia* **33**:349-390.

Steinhaus, E. A., and M. E. Martignoni. 1970. *An Abridged Glossary of Terms Used in Invertebrate Pathology,* 2nd Ed. Pacific Northwest Forest and Range Expt. Sta., USDA Forest Service. 38 pp.

Summers, M., R. Engler, L. A. Falcon, and P. Vail. 1975. Baculoviruses for insect pest control: Safety considerations. Selected papers from EPA-USDA working symposium. *Am. Soc. Microbiol.* 186 pp.

Sussman, A. S. 1951. Studies of an insect mycosis. I. Etiology of the disease. *Mycologia* **43**:338-350.

Swellengrebel, N. H. 1919. *Myiobium myzomyiae* n.g., n. sp., een parasitische Haplosporidie uit het darmkanaal van eenige Anophelinen. *Mededeel. Burgerlijk. Geneeskund. Dienst Ned.-Indie* **10**:68-72.

Thomas, G. M. 1974. Diagnostic techniques. In: *Insect Diseases,* Vol. I. (Ed. G. E. Cantwell). Marcel Dekker, Inc., New York. 1-48.

Thomas, G. M., and G. O. Poinar, Jr. 1973. Report of diagnoses of diseased insects 1962-1972. *Hilgardia* **42**:261-360.

Thomson, H. M. 1960. A list and brief description of the Microsporidia infecting insects. *J. Insect Pathol.* **2**:346-385.

Tuzet, O., J. Maurand, A. Fize, R. Michel, and B. Fenwick. 1971. Proposition d'un nouveau cadre systematique pour les genres de microsporidies. *C. R. Acad. Sci.* **272**:1268-1271.

Umphlett, C. J., and C. S. Huang. 1972. Experimental infection of mosquito larvae by a species of the aquatic fungus, *Lagenidium. J. Invertebr. Pathol.* **20**:326-331.

Vaughn, J. L. 1974. Virus and rickettsial diseases. In: *Insect Diseases,* Vol. I. (Ed. G. E. Cantwell). Marcel Dekker, Inc., New York. 49-85.

Veen, K. H., and P. Ferron. 1966. A selective medium for the isolation of *Beauveria tenella* and of *Metarrhizium anisopliae. J. Invertebr. Pathol.* 268-269.

Veremtchuk, G. V., and I. V. Issi. 1970. On the development of insect microsporidians in the entomopathogenic nematode, *Neoaplectana agriotos* (Nematodes: Steinernematidae). *Parasitologiya* **4**:3–7. (In Russian)

Vincent, M. 1927. On *Legerella hydropori* n. sp. a coccidian parasite of the malpighian tubules of *Hydroporus palustris* L. (Coleoptera). *Parasitology* **19**:394–400.

Wallace, F. G. 1966. The trypanosomatid parasites of insects and arachnids. *Exp. Parasitol.* **18**:124–193.

Waterhouse, G. M. 1973. Entomophthorales. In: *The Fungi, an Advanced Treatise* (Eds. G. C. Ainsworth, F. K. Sparrow, and A. S. Sussman). Academic Press, New York. 219–229.

Weinman, D. 1974. Family II Bartonellaceae. In: *Bergey's Manual of Determinative Bacteriology*, 8th Ed. (Eds. R. E. Buchanan and N. E. Gibbons). Williams & Wilkins Co., Baltimore. 903–906.

Weiser, J. 1961. Microsporidia as Parasites of Insects. Monog. zur Angew. Entom. Beihefte zur *Zeitschrift für Angew. Entomol.* **17**:1–149.

Weiser, J., and J. D. Briggs. 1971. Identification of pathogens. In: *Microbial Control of Insects and Mites* (Eds. H. D. Burges and N. W. Hussey). Academic Press, New York. 13–66.

Weiss, E. 1974. Tribe III Wolbachieae, genus VII *Wolbachia,* and genus X *Rickettsiella.* In: *Bergey's Manual of Determinative Bacteriology,* 8th Ed. (Eds. R. E. Buchanan and N. E. Gibbons). Williams & Wilkins Co., Baltimore. 897–903.

Weiss, E., and J. W. Moulder. 1974. Genus I *Rickettsia,* genus II *Rochalimaea,* and genus III *Coxiella.* In: *Bergey's Manual of Determinative Bacteriology,* 8th Ed. (Eds. R. E. Buchanan and N. E. Gibbons). Williams & Wilkins Co., Baltimore. 883–893.

Whistler, H., S. L. Zebold, and J. A. Shemanchuk. 1974. Alternate host for mosquito-parasite *Coelomomyces. Nature* **251**:715–716.

Whittaker, R. H. 1969. New concepts of kingdoms of organisms. *Science* **163**:150–160.

Wildy, P. 1971. Classification and nomenclature of viruses. *Monographs in Virology* **5**:81 pp.

Wille, H. 1956. *Bacillus fribourgensis,* n. sp., Erreger einer "milky disease" im Engerling von *Melolontha melolontha* L. *Mitt. Schweiz. Entomol. Ges.* **29**:271–282.

Woolever, P. 1966. Life history and electron microscopy of a Haplosporidian, *Nephridiophaga blatellae* (Crawley) n. comb. in the malpighian tubules of the german cockroach, *Blatella germanica* (L.). *J. Protozool.* **13**:622–642.

Wyss, C. 1974. Sporulationsversuche mit drei Varietäten von *Bacillus popilliae* Dutky. *Zentralblatt Bakt., Parasitenk. Infektionskr. Hygiene* **126**:461–492.

Yarwood, E. A. 1937. The life cycle of *Adelina cryptocerci* sp. nov., a coccidian parasite of the roach *Cryptocercus punctulatus. Parasitology* **29**:370–390.

Zacharuk, R. Y., and R. D. Tinline. 1968. Pathogenicity of *Metarrhizium anisopliae,* and other fungi for five elaterids (Coleoptera) in Saskatchewan. *J. Invertebr. Pathol.* **12**:294–309.

INDEX